THE GENERAL THEORY OF SORPTION DYNAMICS AND CHROMATOGRAPHY

VVEDENIE V OBSHCHUYU TEORIYU DINAMIKI SORBTSII I KHROMATOGRAFII

ВВЕДЕНИЕ В ОБЩУЮ ТЕОРИЮ ДИНАМИКИ СОРБЦИИ И ХРОМАТОГРАФИИ

THE GENERAL THEORY OF
SORPTION DYNAMICS
AND
CHROMATOGRAPHY

by Vladimir Vatslavovich Rachinskii

Authorized translation from the Russian

CONSULTANTS BUREAU
NEW YORK
1965

ISBN 978-1-4757-0063-3   ISBN 978-1-4757-0061-9 (eBook)
DOI 10.1007/978-1-4757-0061-9

The Russian text was published by Nauka in Moscow in 1964 for the
Institute of Physical Chemistry of the Academy of Sciences of the USSR.

Введение в общую теорию
динамики сорбции и хроматографии

Рачинский Владимир Вацлавович

Library of Congress Catalog Card Number 65-15004
© *1965 Consultants Bureau Enterprises, Inc.*
Softcover reprint of the hardcover 1st edition 1965
227 West 17th St., New York, N.Y. 10011

PREFACE

The chromatographic method of the analysis and separation of mixtures of substances has been extremely vigorously developed in recent years. The regions of its application are constantly expanding. The method of analysis is being improved. Chromatographic methods are finding more and more use in industry. However, it should be recognized that the rate of development of the theory of chromatography is lagging appreciably behind the practice of this method, which is graphically revealed by an acquaintance with books on chromatography published in recent years. Of course, this is not accidental, since actually the theory of chromatographic processes is one of the most difficult theoretical problems. The lag in the field of chromatographic theory is due to a certain degree to the fact that this problem until now had not interested the representatives of physicomathematical sciences, especially mathematics. ·

Various chromatographic processes obey a number of general laws—the laws of the so-called dynamics of sorption. Chromatography is one of the means of practical utilization of the dynamics of sorption.

In this book an attempt was made to present the bases of the general theory of the dynamics of sorption and chromatography in a logical sequence.

The author, being by specialty a physical chemist, definitely recognizes that the book presented to the reader contains certain shortcomings. This is due chiefly to mathematical difficulties that arise in the solution of the problems of the dynamics of sorption and chromatography. Not all the formulations and solutions could be given with exhaustive completeness and rigorousness. Great attention was paid to the elucidation of the physical meaning of the formulation and solution of problems of the dynamics of sorption.

The main purpose of the book was to promote the theoretical training of practical specialists in chromatography. The book does not pretend to any exhaustive coverage of all the general theoretical problems of the dynamics of sorption. It does not discuss particular theories of various types and varieties of chromatography. Individual works should be devoted to them.

The author would like to express his profound gratitude to Professor O. M. Todes, Doctor of Physicomathematical Sciences, for his valuable advice and discussion.

CONTENTS

Chapter I. Dynamics of Sorption and Its Practical Significance   1
    1. Dynamics of Sorption as a Heterogeneous Process   1
    2. On the History of the Practical Applications of the Phenomenon of Sorption Dynamics.
        Sorption Technology and Chromatography   3
    3. Problems of Terminology and Classification in the Theory of the Dynamics
        of Sorption and Chromatography   7
    4. History of the Development of the Theory of Dynamics of Sorption
        and Chromatography   9

Chapter II. General Formulation of the Problem of the Dynamics of Sorption and Methods
        of Its Solution   13
    1. Formulation of the Problem   13
    2. Equations of the Material Balance   14
    3. Equations of the Kinetics and Statics of Sorption   15
    4. Equations of Hydrodynamics   17
    5. Initial and Boundary Conditions   19
    6. Simplification of the System of Equations of Sorption Dynamics   19
    7. Methods of Solving the Problems of Sorption Dynamics   23
    8. Statistical Method   25

Chapter III. Theory of the Frontal Dynamics of Sorption of One Substance   27
    1. Initial Equations   27
    2. Equilibrium Sorption Dynamics in the Absence of Longitudinal Effects ($D^* = 0$)   28
    3. Equilibrium Sorption Dynamics under the Action of Longitudinal Effects ($D^* \neq 0$)   37
    4. Nonequilibrium Sorption Dynamics in the Absence of Longitudinal Effects ($D^* = 0$)   45
    5. Nonequilibrium Sorption Dynamics under the Action of Longitudinal Effects ($D^* \neq 0$)   48

Chapter IV. Theory of Frontal Chromatography   51
    1. Equilibrium Sorption Dynamics of a Mixture of Substances in the Absence
        of Longitudinal Effects ($D^* = 0$)   51
    2. The Frontal Chromatogram   56
    3. Equilibrium Sorption Dynamics of a Mixture of Substances under the Action
        of Longitudinal Effects ($D^* \neq 0$)   64
    4. Nonequilibrium Sorption Dynamics of a Mixture of Substances   66

Chapter V. Theory of Elution Chromatography   69
    1. Elution Sorption Dynamics of One Substance   69
    2. Elution Sorption Dynamics of a Mixture of Substances   73

Chapter VI. Theory of Displacement Chromatography   77
    1. Equilibrium Displacement Sorption Dynamics in the Absence of Longitudinal
        Effects ($D^* = 0$)   77
    2. Displacement Sorption Dynamics under the Action of Longitudinal and Kinetic Effects   82

Literature Cited   85

CHAPTER I

DYNAMICS OF SORPTION AND ITS PRACTICAL SIGNIFICANCE

1. Dynamics of Sorption as a Heterogeneous Process

Let us consider the process of the dynamics of sorption from the general physicochemical standpoint. First of all, let us briefly mention certain thermodynamic and physicochemical concepts [2, 44, 82, 83].

In the analytical approach to the study of the properties and phenomena of the material world, one must distinguish its individual parts and study them separately.

A body or group of interacting bodies, studied separately, is customarily called a system in thermodynamics. Any physically homogeneous body or set of identical and physically homogeneous bodies, contained in the system, is called a phase. Phases consisting of any one chemically individual substance are called simple or pure, while phases containing several individual chemical substances are called mixed.

A system consisting of only one phase is called homogeneous; a system consisting of several phases is called heterogeneous. Among nonuniform particles, phases of a heterogeneous system, there exist surfaces, boundaries of phase separation (interfaces).

The processes that take place in homogeneous systems are called homogeneous, while those that take place in heterogeneous systems are called heterogeneous.

Each phase is characterized at each given moment by a definite thermodynamic state, i.e., by a composition, pressure, volume, temperature, and other physical parameters. Any system as a whole and its individual parts can exist in a state of emergence, change, and relative motion. Heterogeneous systems can arise in different ways. For example, when a gas is cooled to a temperature below the critical point, the homogeneous, one-phase system—the gas—is converted to a heterogeneous liquid-vapor system. Analogously, when a solution of a salt is cooled, at some moment a precipitate begins to form—a solid phase—and the one-phase system is converted to a heterogeneous system of solid salt and saturated solution. A heterogeneous system can also be formed by the combination of several phases into one general system.

A heterogeneous system can exist in an equilibrium or a nonequilibrium state. In the equilibrium state, its composition and thermodynamic parameters remain constant with time. If the parameters of the system experience spontaneous changes in accord with the second principle of thermodynmanics, then such a system will be nonequilibrium. The spontaneous process of change of state of a nonequilibrium system ends in the establishment of an equilibrium state.

Let us assume that a nonequilibrium heterogeneous system has been formed by the combination of several phases. The transition of the system from the nonequilibrium state to the equilibrium state is accomplished as a result of the interaction of the phases of the heterogeneous system that have come into contact.

Spatially, the process of phase interaction begins with a molecular or chemical interaction at the interface. Atoms, molecules, or ions situated at an interface possess increased activity for intermolecular or chemical interaction, as a result of the unsaturation of their intermolecular or chemical bonds. If only intermolecular forces operate in the process of phase interaction, then the chemical composition of the heterogeneous system on the whole is unchanged. Only a change in the chemical composition of the individual phases occurs as a result of the transition of substances from one phase to another. The process of interaction in this case ends in the establishment of a definite equilibrium distribution of substances among the phases of the heterogeneous system. A heterogeneous system in the state of thermodynamic equilibrium will be characterized by a definite equilibrium composition of the phases. If the forces of a chemical bond participate in the interaction, on the other hand, and, as a result, chemical reactions develop, a change occurs in the chemical composition both of the heterogeneous system as a whole and of its individual phases. In this case, profound changes can occur in the heterogeneous system, involving the formation of new chemical compounds and new phases. The process of chemical interaction of the phases of a heterogeneous system ends in the establishment of a definite equilibrium distribution of substances between the original and newly formed phases. The phases of the heterogeneous system will also be characterized by a definite equilibrium composition.

The transition of a substance to a given phase from another phase in the process of establishment of interphase distribution of the substances can be considered as a process of absorption or sorption of substances by the given phase. Since the substances of the system are distributed to one degree or another among all the phases of the heterogeneous system, the sorption process is carried out in each of the phases. In principle, each of the phases of a heterogeneous system is a sorbent, i.e., a medium that absorbs substances.

The course and character of the interaction of the phases of a heterogeneous system, and, consequently, the course and character of the sorption processes, depend on the state of aggregation of the phases, properties and composition of all the components of the system.

From the standpoint of the spatial distribution of the substances to be sorbed, we can indicate two types of sorption: adsorption and absorption. Adsorption is a sorption process in which the substances to be sorbed are concentrated only at the interface (surface). Absorption is a sorption process in which the substances to be sorbed are distributed in the volume of the interacting phases. The phenomena of adsorption occur, as a rule, when solid phases participate in the sorption process. Steric, geometrical conditions, hindering diffusion penetration of the substance to be sorbed into the solid phase, serve as an obstacle to the extension of the sorption process into the volume of the entire solid phase [17]. However, we should keep in mind that even in the case of absorption, uniform distribution of the substance to be sorbed in the volume of the phases of the heterogeneous system does not always occur. In view of the increased sorption activity of the boundary layers of the phases, the substances to be sorbed can be concentrated to a great degree at the interface. In certain cases, diffusion layers with concentrations of the substances to be sorbed gradually varying within the phases are formed close to the interface.

In polycomponent heterogeneous systems, during their approach to a stable thermodynamic state, extremely varied distributions of substances between the phases in contact can arise. In certain cases, as a result of atomic-molecular interactions at the interface, which take place with different activities for different substances, the individual components of the system not only do not pass from one phase into the other, but, on the contrary, are displaced by other surface-active substances, back into the same phase, and the relative concentration of such a component in the volume of this phase does not decrease, but increases. Such a phenomenon is called negative sorption.

Two basic types of forces of atomic-molecular interaction operate in the process of phase interaction: intermolecular (Van der Waals bond) and chemical (chemical bond). The molecular bond possesses three basic varieties: orientational, induction, and dispersion. The chemical bond possesses two basic varieties: heteropolar (ionic) and homeopolar (covalent) bonds. Three basic types of sorption can be distinguished according to the nature of the forces of atomic-molecular interaction: molecular, heteropolar (ionic), and homeopolar (covalent) [30, 34, 107, 111]. Heteropolar and homeopolar sorptions can be called by the single common term of chemical sorption.

Since the boundary layers possess the greatest activity in phase interaction, in practice it is expedient to create the greatest area of the surface of phase separation to accomplish the most effective sorption of substances in the quantitative and kinetic respects (high rate of establishment of sorption equilibrium). This is achieved by dispersion of the phases.

In nature and in technological sorption processes, the interaction of the phases of a heterogeneous system can occur both under conditions of their relative rest and under conditions of relative motion. The character of the relative motion of interacting phases can be dual: random, chaotic mixing of the phases, or their directed relative motion. In the case of relative rest or random mixing, the so-called static sorption occurs, while in the case of directed relative motion, dynamic sorption occurs. Let us give more complete definitions of these concepts.

Static sorption (statics of sorption) represents a sorption process that takes place in the presence of relative rest or random mechanical mixing of the phases of a heterogeneous system and ends in the establishment of sorption equilibrium among the interacting phases. Dynamic sorption (dynamics of sorption) represents a sorption process accomplished under conditions of directed relative motion of the interacting phases. The setup of sorption experiments or the technological utilization of sorption is accomplished in two basic variations: static and dynamic sorption.

The processes of sorption statics and dynamics have long found practical application in sorption techniques. The phenomenon of sorption dynamics lies at the basis of chromatographic methods of separation of mixtures of substances. Any natural phenomena in the nonliving and living world are accomplished on the basis of the laws of sorption dynamics. Chromatography has now acquired the significance of one of the basic practical methods of utilizing the phenomena of sorption dynamics. Hence, a theoretical investigation of sorption dynamics is essentially at the same time a development of the theory of chromatography as well.

During the long historical period, the technical and analytical applications of sorption dynamics have been developed, as it were, independently, and yet, they possess a common theoretical base [107]. Only recently has the consolidation of these trends on the basis of a common theory begun.

The modern state of the theory and practice of sorption technology and chromatography is expediently considered in its historical relationship to the general development of the theory of sorption processes.

2. On the History of the Practical Applications of the Phenomenon of Sorption Dynamics. Sorption Technology and Chromatography

It is clear from the literature sources [79, 130, 159, 214] that the sorption of substances by porous media and the dynamic method of carrying out the sorption process were already known in ancient times. In the works of Aristotle [157, 159], a description of the production of fresh water from sea water by filtration through a layer of soil or sand is encountered. The further many-century development of the sorption technique proceeded along the line of the testing and application of various natural materials, chiefly for the purification of water.

A new stage in the development of the theory of sorption phenomena and the sorption technique began in the era of the development of capitalism. In 1733, the Swede K. Scheele and in 1777, the Frenchman A. Fontana independently discovered the high sorption capacity of wood charcoal with respect to gases [44, 47, 206, 209]. However, while this discovery did not find practical application immediately, still, it did play a definite role in the development of the kinetic theory of gases. In 1785, the Russian academician T. E. Lovits [88, 124] discovered the sorption properties of charcoal with respect to dissolved substances. This discovery immediately found practical application. Charcoal began to be used in the purification of water, tartrate, acetic acid, grain alcohol, in the clarification of syrup in sugar production, for the refinement of plant oils, saps, etc. In addition to the static method of using charcoal as a sorbent, charcoal sorption filters, also proposed by T. E. Lovits [88], were also successfully used.

The further development of the theory of sorption processes was related to studies in the field of the scientific bases of agriculture. A number of new phenomena were discovered in the investigation of the sorption capacity of soils and minerals. The German scientist J. Liebig [204, 205] revealed that in the filtration of a mixture of salts through soils, potassium is retained in the upper layers, while sodium is retained in the lower layers. He ascribed this different distribution of the two elements to the different sorption of potassium and sodium on soil particles. T. Clark used filters of clay minerals for the softening of water. He found that clay possesses an ability to absorb calcium and magnesium from solution [206].

During the period from approximately 1845 to 1855, H. Thompson and I. Way [79, 231], J. Liebig [204, 205], and other scientists conducted extensive investigations of the sorption of salts and organic substances of manure by soils. I. Way made an important contribution to the experimental proof of the existence of exchange sorption of substances [231]. In 1858, E. Eichgorn [179] showed that the phenomenon of exchange sorption of mineral substances on chabazite and natrolite is reversible in character. The concept that the exchange sorption of mineral substances is based on the phenomenon of ion exchange grew up somewhat later in connection with the development of the theory of electrolytic dissociation and the theory of atomic structure. With the discovery of the phenomenon of exchange sorption in soils and minerals, the possibilities of the sorption technique were substantially expanded. Sorption filters, filled with clay and silicate minerals, began to be used in industry.

S. Poggendorf [215] and I. Harm [200] proposed the use of silicates for the removal of potassium from sugar solutions. A great contribution to the use of ion exchange methods in industry was made by the German scientist R. Gans, who received a number of patents in 1905-1915 for the invention of installations for the softening of water, purification of sugar solutions and other liquids [183].

In the industry of Russia, zeolite filters were also successfully used at the sugar plants [125]. Ion exchange filters were also used for the softening of water [16, 64].

The technique of sorption filters also found wide use in the petroleum industry. At approximately the same time, sorption filters (limestone, clays) were used for the fractionation of petroleum in Russia by S. K. Kvitka [75, 78], in Germany by C. Engler and E. Albrecht [181], and in the Unites States by D. Day [178] and by J. Gilpin [184, 185].

During the first World War, in connection with the use of poison gases at the front, a new branch of sorption technique arose—antichemical protection. In 1915, the Russian scientist N. D. Zelinskii invented a gas mask in which a charcoal filter was used as a sorbent for poison gases and vapors [62].

We should mention specially that the most important successes in the field of the use of the sorption technique were related to a considerable degree to the use of new sorption materials. Great efforts were applied to the production of materials with high sorption activity and capacity. T. E. Lovits [88] was the first to discover that the sorption capacity of charcoal can be considerably increased by special treatment (for example, with sulfuric acid, by boiling). Comprehensive investigations of the activation of charcoals were conducted in 1915-1917 by N. D. Zelinskii [44, 47, 62].

In connection with the development of means of chemical protection, both in Russia and in other countries, work was expanded on the investigation of the sorption of gases and vapors by charcoals and other sorbents [44, 47]. In a number of countries whole schools arose, working along this line. In our country, the school of N. A. Shilov, which made a great contribution to the theory of gas sorption [85, 168], achieved renown. The results of the work of this school are outlined in detail in the monographs of M. M. Dubinin and K. V. Chmutov [44, 47].

In addition to chemical protection, charcoal filters have also begun to be used successfully in chemical industry for the regeneration and removal of gases and vapors, especially in the petroleum industry [44, 132]. A technology of charcoal filters for the production of water and solutions has also been developed [24].

The low exchange capacity of natural minerals, used for the preparation of ion exchange sorption filters, was the cause of searches for synthetic inorganic sorbents. Chemists succeeded in producing synthetic aluminosilicates, which possessed many times greater sorption capacity than natural silicates [183].

In the history of the theory of sorption processes, the works of K. K. Gedroits [38] were of great practical and theoretical significance. He and other researchers discovered that soil humus possesses considerably greater sorption capacity than its mineral portion. The concept that the inorganic portion of the soil possesses the greatest sorption capacity subsequently led to the development of new methods of producing sorbents with high exchange capacity. The first great achievement in this field was the preparation of the so-called sulfonated coals, produced by the humification of coal and peat. A second, decisive step was the development of the synthesis of high-molecular organic sorbents—synthetic ion exchange resins [4, 79, 127, 170, 201].

During the last decades, great work has been done in the Soviet Union in the field of the further development and introduction of sorption technology into various branches of the national economy: chemical, food, medical industry, water preparation, etc. [4, 68-73, 126, 127, 135, 156]. The Soviet scientists have developed a technology for the production of high-quality sorbents—activated charcoals, silica gels, alumina gels, permutits, and ion exchange resins [4, 68-73, 103, 104, 126, 127, 135, 156].

However, the history of the technical application of the dynamics of sorption, as we have seen, shows that sorption technology has been limited chiefly to problems of simple purification of substances. Moreover, as a rule, the simplest variation of dynamic sorption has been used—the technique of the sorption filter. Individual attempts (S. K. Kvitka, D. Day, J. Gilpin, C. Engler, and E. Albrecht) to use sorption columns for the industrial separation of complex mixtures of substances did not lead at that time to the development of an independent sorption method of separating mixtures. Industrial methods of separating mixtures of substances were constructed (and are still constructed even now) chiefly on the basis of such physical methods as distillation, rectification, repeated recrystallization, sublimation, flotation, thermal diffusion, and others, as well as on various purely chemical methods of separating mixtures. This is explained by the fact that the problem of fine and complete separation of complex mixtures of substances in industry was not so acute before.

An entirely different situation has arisen in analytical chemistry, especially in analytical biochemistry. In this field, an acute need for simple and fine methods of separation and analysis of complex mixtures has always existed. It should be mentioned that attempts to use dispersed materials for analytical purposes also date back to ancient times. Thus, for example, there is a mention in the literature [9] that Pliny used papyrus impregnated with an extract of gall nuts for the detection of iron sulfate. In the middle of the 19th century, investigations appeared devoted to the use of paper for analytical purposes. Of the earliest works, we might mention the works of A. Cohn [176], who detected the formation of colored rings in the evaporation of a drop of a solution of plant pigments on paper, as well as the works of F. Runge [166, 216], who studied periodic processes in the flow of solutions of salts and alkalies along paper, preliminarily impregnated with various precipitants. Of definite historical significance in the development of these initial works were the investigations of F. Schönbein [217], and especially F. Göppelsröder [198], who was the creator of the so-called capillary analysis.

Although the indicated researchers did not know the true nature of the processes that occur in the motion of substances in a porous medium of paper, however, actually these were the first and important attempts to use the dynamics of sorption for analytical purposes. In spite of certain successes of capillary analysis, its potentialities were extremely limited. Considerably later, already in our time, filter paper began to be used for conducting so-called spot test, developed by N. A. Tananaev [134] and F. Feigl [151], as well as electrocapillary analysis, developed by S. L D'yachkovskii [48-50].

However, neither drop reactions nor simple capillary absorption of solutions solved one of the most important problems of analytical chemistry, the complete separation of a mixture of substances, followed by their analytical investigation.

This problem was brilliantly solved by the Russian scientist M. S. Tsvet, who in 1903 discovered the chromatographic method of separating mixtures of substances [158]. M. S. Tsvet, being a botanist, devoted his scientific activity to the investigation of plant pigments. However, he was interested in the question of the nature and physiological role of chlorophyll. Embarking upon a study of this difficult and important problem, M. S. Tsvet encountered a need for developing a reliable and effective method of separating and isolating plant pigments in pure form. During his work on a method for the extraction of pigments from plant leaves with various solvents, he noted that chlorophyll is poorly extracted when the leaves are treated with gasoline or petroleum ether. This phenomenon could not be explained by the poor solubility of chlorophyll in these solvents, since it was known that chlorophyll dissolves in them rather well. Analyzing the causes of this phenomenon, M. S. Tsvet established that chlorophyll exists in plant leaves in the sorbed state, and does not pass into solution when treated with petroleum ether, since petroleum ether as a solvent cannot desorb chlorophyll bound to the colloidal substances of the chloroplasts of plant cells. He made a systematic investigation of the adsorption of pigments on various sorbents, attempting to find practical ways to utilize adsorption phenomena for the separation of pigments. It should be mentioned that at the time when M. S. Tsvet conducted his investigations (1899-1914), adsorbents were already in wide use in science, especially in technology. As was already noted above, two methods of using adsorbents for the purification of substances were known—under static and under dynamic conditions. Thus, in his investigations, M. S. Tsvet was already able to rest on definite experience of science and technology on the use of absorbents.

He showed in numerous experiments that chlorophyll is adsorbed from petroleum ether by very many substances of organic and inorganic character. M. S. Tsvet also established that the pigments of plant leaves possess different abilities to be adsorbed on various adsorbents in the presence of various solvents. He used this important experimental fact as the basis for further searches for new and effective methods of separating pigments.

Differences in adsorbability were successfully utilized in biochemical investigations only once before M. S. Tsvet—by the Russian scientist A. I. Danilevskii [40] for the separation of amylase from trypsin.

5

M. S. Tsvet [158] tested two methods of utilizing the differences in adsorbability for the separation of pigments—under static and under dynamic conditions. He showed that the dynamic method is the most effective. According to this method, a solution containing a mixture of pigments is passed through a layer of adsorbent. As a result of differences in adsorbability, the individual pigments are arranged in the column of the adsorbent in the form of individual, differently colored zones; when the column is eluted with a pure solvent, the zones, moving along the column at different rates, become more and more individualized, and, finally, they are completely separated, each zone containing only one pigment. M. S. Tsvet called this new method of separating mixtures of substances chromatography. He detailed the bases of the chromatographic method, as well as the results of investigations of plant pigments by this method, in the classic work "Chromophylls in the Plant and Animal World," published in 1910 [159].

M. S. Tsvet's scientific merit also lies in the fact that he not only developed a new method of separating and analyzing mixtures of substances, but also correctly foresaw the significance and prospects of its development. The mechanism of dynamic molecular adsorption was utilized in the classical chromatographic method for the separation of mixtures of substances. However, the works of M. S. Tsvet also contain indications that the separation of substances under dynamic conditions may occur as a result of different physicochemical mechanisms of the sorption process: sorption of ions, formation of precipitates, etc.

The significance of the chromatographic method was underestimated, both during the lifetime of M. S. Tsvet, and subsequently. Only in the thirties did conditions arise for the regeneration of this method and its further development. From the field of biochemistry, the chromatographic method began to penetrate rapidly into organic, inorganic chemistry, chemical technology, and other branches of science and technology [107]. M. S. Tsvet's predictions were realized. New types and varieties of the chromatographic method were discovered and began to be developed. G. Schwab (1937-1940) [219] was the founder of ion exchange chromatography, which received further development in the works of the Soviet scientist E. N. Gapon (1947-1950) [26, 30, 34-36].

In 1941, Martin and Synge [207] developed a new variety of molecular chromatography—partition chromatography. Somewhat later (in 1944), R. Consden and other researchers [177] proposed a method of producing partition chromatograms on paper. At the present time paper chromatography is one of the widespread analytical methods for complex mixtures [157].

E. N. Gapon et al. discovered the method of precipitation chromatography of ions in 1948 [5, 28, 37].

Together with the development of chromatography from solutions, the chromatography of gases and vapors, also widely used in recent years in chemistry and biology, was also developed. M. M. Dubinin and M. V. Khrenova [45] were the first to produce a chromatogram of vapors in 1936. A. A. Zhukhovitskii and N. M. Turkel'-taub et al. developed new variations of gas and vapor chromatography—chromathermography, the thermodynamic method, partition gas chromatography [18, 54-60]. The studies of A. Tiselius and S. Claesson [228, 77], who proposed a number of methods for the analysis of complex mixtures on the basis of the production of chromatographic "effluent curves," are of great significance in the improvement of chromatographic methods.

The chromatographic method, called to life by the practical needs of biology, is now playing a progressive role in the investigation of complex processes of metabolism in living organisms [107, 128, 157, 213].

The complication of problems of industrial production, involving the necessity of conducting a complete separation and analysis of complex mixtures (natural raw materials and industrial products), is the cause of the recently observed introduction of the chromatographic method of separating mixtures into various branches of industry [4, 67, 76, 79, 97, 152, 155, 199, 203]. The use of sorption filters is now recognized as a specific technique—the simplest case of chromatographic separation as a higher form of the sorption technique.

Thus, the modern stage in the development of the practical application of the dynamics of sorption is characterized by a combination of the technical and analytical trends in its application. This combination is accompanied by a leading role of chromatography. Since this book is devoted exclusively to general problems of the theory of the dynamics of sorption and chromatography, we shall cite in the list of literature only works treating the practical bases of the sorption technique and chromatography [4, 67-73, 76, 77, 79, 97, 107, 126-129, 132, 152, 153-157, 161, 167, 173-175, 199, 203, 213, 238, 240].

In the construction of the theory of the dynamics of sorption and chromatography, questions of terminology and classification of the processes and methods of work are of considerable importance. The setting of the terms and concepts in a definite order, the creation of a definite classification of the processes is the first and essential step in the theoretical generalization of any knowledge. At the beginning of this chapter, we cited certain theoretical concepts characterizing the physical nature of the phenomenon under study—the dynamics of sorption. Here we shall continue our consideration of the problems of terminology and classification in modern chromatography. Of course, we should make the reservation that any terminology and classification is to one degree or another arbitrary. But nonetheless, it should be sufficiently theoretically substantiated, i.e., strictly scientific. From this standpoint, we must first of all consider in greater detail the question of the basic essence of the chromatographic method in its modern meaning.

This problem is of significance not only for the creation of a rational classification of types and varieties of chromatography, but also for the prediction of ways of further development of the method.

As has already been mentioned, M. S. Tsvet developed the chromatographic method of separating mixtures on the basis of an observation of the different adsorbability of plant pigments. The investigation of this phenomenon permitted him to formulate the following law, which he called the law of adsorption replacement [159, 160]: "Substances dissolved in a definite liquid form a definite adsorption series A, B, C, . . . , expressing the relative adsorption affinity of its members for the adsorbent. Each of the members for the adsorption series, possessing greater adsorption affinity than the following member, displaces it from the compound and in turn is displaced by the preceding member." M. S. Tsvet also formulated the basic necessary condition for the separation of substances by adsorption methods. He wrote: "In order for two substances existing in solution to be able to be separated according to adsorption methods, they must occupy different ranks in the adsorption series." In the description of the physical essence of the chromatographic process, M. S. Tsvet concludes, on the basis of the law of adsorption displacement: "The zonal distribution of the components of a system [in a chromatographic column—V.V.R.] expresses the relative position of the latter in the adsorption series."

The premises indicated above, formulated by M. S. Tsvet, were at that time sufficient for a theoretical substantiation of the chromatographic method. However, it should be kept in mind that, as has already been noted, M. S. Tsvet used only one type of sorption—molecular—in his method. Modern chromatography, on the other hand, uses a great variety of sorption mechanisms, and in view of this, a more general and broader understanding of the essence of the chromatographic method is needed.

It is difficult to give an all-encompassing formulation of the basic principle of chromatography at the modern stage of its development. Attempts to give such a formulation [107, 111, 239, 240] have evoked a number of objections [145, 157]. However, the disclosure of the most general and principal features of the chromatographic method is one of the primary problems of the theory of the dynamics of sorption and chromatography.

It is expedient to mention these most general features of the chromatographic method. It should be considered as a practical utilization of the physical phenomenon of the dynamics of sorption. The method is designed for purposes of concentration, purification, and separation of mixtures. The separation of substances during the process of dynamic sorption occurs as a result of differences in the sorption distribution of the substances between the stationary and mobile phases of the heterogeneous system.

Any mechanism of sorption—interphase distribution—can be used for the chromatographic separation of substances.

The first attempt to give a classification of types of chromatography, types of chromatograms, and methods of their production and analysis was made by E. N. Gapon and T. B. Gapon [27, 29, 30, 34]. They used the character of the interaction between the substances to be separated and the sorbent as the basis for the classification of types of chromatography.

Concepts of two basic types of sorption—molecular sorption and chemosorption—were given above in accord with the nature of the atomic-molecular interaction. A classification of types of chromatography can also be given according to the same feature, distinguishing two basic types—molecular and chemosorption chromatography. Each type may contain its own varieties, characterized both by the specific peculiarities of the operating forces of atomic-molecular interaction and by other features.

Thus, for example, at the present time two varieties can be distinguished in molecular chromatography: adsorption molecular chromatography (the classical variation of the chromatographic method of M. S. Tsvet) and absorption (partition) molecular chromatography, proposed by A. Martin and P. Synge. In adsorption molecular chromatography, the sorbent is a solid dispersed material. The sorption process occurs on the surface of the solid phase. In absorption (partition) molecular chromatography, the role of the unique sorbent (absorbent) is played by a liquid phase, which is retained stationary on the surface of a solid phase (carrier) by sorption forces of intermolecular character. The formation of adsorption molecular chromatograms occurs as a result of quantitative differences in the adsorbability of the substances, while in absorption chromatography, it occurs as a result of differences in the distribution coefficients between two immiscible liquid phases, one of which is stationary, while the other is mobile. However, in both cases the same forces operate—forces of intermolecular bonding, on the energy of which the character of the distribution of the substances to be chromatographed between the bounding and interacting phases depends.

The following well-known types of polar sorption can be indicated: sorption of potential-determining ions, ion exchange and precipitation sorption of ions. The sorption of potential-determining ions plays an important role in processes of overcharging of ionogenic sorbents, as well as in the phenomenon of the secondary sorption of ions on certain ion exchange sorbents.

Differences in the exchange sorption of ions are utilized for the production of ion exchange chromatograms. The corresponding variety of chemosorption chromatography is called ion exchange chromatography.

Precipitation sorption of ions here subsumes the process of formation of ionic crystalline precipitates in the chromatographic column. This type of polar sorption has been utilized in the method of precipitation chromatography, proposed by E. N. Gapon.

The process of homeopolar sorption is a chemical process of interaction, related to the manifestation of covalent forces. Here the most varied cases are possible, beginning with profound chemical reactions leading to complete breakdown of the previous structure of the sorbent and the formation of a new solid phase, and ending with various types of surface chemical reactions; in particular, the so-called activated adsorption belongs to this type of sorption [1, 14, 86, 87, 103, 106, 123]. The corresponding variety of chemosorption chromatography is quite new [39].

Differences in the sorbability of substances can be extremely varied. These differences are due to a different type of sorption (intertype differences). For example, two substances can be chromatographically separated under the condition that one of them is sorbed according to the molecular type, while the second is sorbed according to the ion exchange type of sorption. Differences in the sorbability of substances in a given type of sorption (intratype differences) can be quantitatively characterized by the corresponding bond energies or by the sorption constants, contained in the equation of the sorption isotherms. For example, in adsorption molecular chromatography, substances are separated as a result of differences in the sorption constants, entering into the equation of the linear isotherms or Langmuir isotherms; in partition chromatography, separation occurs as a result of differences in the partition coefficients in the partition equation; in ion exchange chromatography as a result of differences in the exchange constants, contained in the equation for the ion exchange isotherms; in precipitation chromatography of ions—as a result of differences in solubility.

In a given type of sorption interaction, differences in the sorption may also be due to structural-geometrical factors. The influence of the structure of the sorbent and structure of the substances to be sorbed on the course of the sorption process is extremely great. This influence is manifested in the fact that, depending on the relative dimensions of the inner pores of the sorbent grains and the dimensions of the molecules or ions to be sorbed, sorption can occur according to the type of extramicellary or intramicellary sorption. From this standpoint, sorptions with internal porosity can be considered as unique "sieves," capable of sorting and separately sorbing molecules or ions of various dimensions.

It should be considered that in a chromatographic column, the most varied sorption processes, as well as accompanying processes of a nonsorption character, frequently occur simultaneously. Nevertheless, experimental conditions can be created, such that the substances will be sorbed only according to one type, characterized by a definite sorption equation; in this case, chromatographic separation of the substances is possible only under the condition of the presence of quantitative differences in the sorption constants for the substances to be separated.

Thus, the possibilities for accomplishing chromatographic processes of separation of mixtures are virtually unlimited.

The use of the most varied heterogeneous processes of interaction of substances for purposes of chromatography is the way for further development of chromatography, a way that will lead to the discovery of new varieties of the chromatographic method.

The chromatographic process is accomplished in practice by the performance of individual operations, which are called methods of chromatography.

Three basic methods of chromatography may be distinguished:

1. Filtration of the initial mixture of substances through a column of a sorbent, resulting in the production of the so-called primary or frontal chromatogram; this method is called frontal chromatography;
2. Washing the column of sorbent with a stream of pure solvent after obtaining the primary chromatogram, resulting in the production of the so-called elution chromatogram; this method is called elution chromatography;
3. Filtration through the column after the production of the primary chromatogram of a solution of some substance (displacing agent), resulting in the production of a displacement chromatogram; the corresponding method is called displacement chromatography.

In addition to these methods, certain supplementary chemical or physical agents which exert an active and sometimes decisive influence on the course of the chromatographic separation of mixtures can be used in chromatography. The chemical agents include, for example, passage of a solution of a complexing agent through the column after production of the primary ion exchange chromatogram (complex-forming elution).

The supplementary physical agents, which sometimes lead to an effective separation of mixtures, include, for example, the use of an electric field (electrochromatography) or thermal influence (thermochromatography). Obviously such possibilities of active influence on the chromatographic process by other factors are as yet far from exchausted.

4. History of the Development of the Theory of Dynamics of Sorption and Chromatography

For a long time, the concepts of the physical nature of the process of the dynamics of sorption, in spite of the great historical antiquity of practical utilization, were extremely primitive. Thus, for example, Aristotle believed that the absorption of salts in the filtration of sea water through soil occurs as a result of the "gravity" of the salts [130, 159]. T. E. Lovits used the phlogiston theory to explain sorption phenomena [124]. F. Göppelsröder believed that the formation of zones of colored substances in the capillarization of substances on paper is a result of the different rate of diffusion of the substances along the capillaries [198]. M. Faraday was the first to correctly consider the sorption process as a manifestation of electric forces. As physics and chemistry, atomic theory, thermodynamics and electrodynamics, and theory of intermolecular and chemical bonds developed in the 19th century, the nature of sorption processes gradually began to be elucidated.

The modern theory of sorption is based on such general physical theories as thermodynamics, statistical physics, electrodynamics, and quantum mechanics.

Our task does not include a detailed elucidation of the modern state of the theoretical concepts of the nature of sorption interactions. This question has been treated in a number of books, monographs, and surveys [1, 4, 10, 14, 23-25, 39, 44, 47, 51, 66, 79, 90, 106, 107, 132, 157, 160, 206].

The physical essence of the dynamics of the sorption of substances was not immediately understood. The Russian engineer S. K. Kvitka in technology and M. S. Tsvet in analytical biochemistry were the first to correctly understand the true physical cause of the different rate of motion of substances during their passage through a layer of a porous material. It lies in the different adsorbability of the substances by the material of the sorption column. Thanks to the use of correct physical representations, M. S. Tsvet was able to establish a number of qualitative principles of the dynamics of sorption, of which we spoke above. Among these principles, the law of sorption displacement is of the greatest significance. M. S. Tsvet also noted the dependence of the course of the formation of chromatograms on the constants of the sorption isotherms. Using the equation of the Freundlich isotherm, popular at that time, he qualitatively predicted the formation of stationary sorption fronts of pigments.

The first quantitative principle of the dynamics of sorption was established by N. A. Shilov et al. [85, 168]. On the basis of a generalization of the experimental data obtained in a study of the dynamics of molecular sorption of gases and vapors, it was determined that the process of dynamic sorption of gases and vapors on activated charcoal consists of two steps: the step of formation of the front and the step of its parallel transport. He proposed the well-known empirical formula of the "protective action" of the sorption filter, which expresses one of the most important principles of the dynamics of sorption.

The studies of N. A. Shilov on the quantitative principles of the dynamics of sorption of gases and vapors were further developed by M. M. Dubinin et al. [41-47].

S. A. Voznesenskii [24] applied N. A. Shilov's empirical formula to the dynamics of the sorption of substances from solutions.

Subsequently, it was necessary to impart a rigorous form of quantitative relationships, following from more general physical laws—the laws of conservation of matter and energy, laws of kinetics and statics of sorption, laws of hydrodynamics, etc.—to the established qualitative and empirical principles.

The first attempts to give a theoretical interpretation of the formula of the "protective action" of the sorption filter was that of the Czechoslovakian scientists W. Mecklenburg and P. Kubelka [93, 208]. They considered the statics of the sorption process from the standpoint of the theory of capillary condensation, and the kinetics from the standpoint of the representations of the diffusion mechanism of sorption. The theory of W. Mecklenburg and P. Kubelka was soon criticized by A. A. Zhukhovitskii et al. [52]. The basic objections pertained to the mechanism of the sorption process on activated charcoals, as well as mathematical inaccuracies. In spite of the definite correctness of this criticism, it still should be recognized that the theory of Mecklenburg and Kubelka generally correctly reflects the dependence of the process of formation of the front on the kinetic and diffusion parameters. In this sense, it played a positive role. A more rigorous general formulation of the problem of the dynamics of sorption, from the physical and mathematical standpoint, was constructed by Ya. B. Zel'dovich [63] and by E. Wicke [233-237]. They were the first to correctly compile a system of differential equations in partial derivatives, describing the dynamics of the sorption process of one substance. The first of these equations reflected the general law of conservation of matter, the second the equation of the sorption isotherm. E. Wicke formulated the initial differential equations for the problem of the so-called equilibrium dynamics of sorption. Solving the problem of the equilibrium dynamics of sorption in the case of a linear isotherm, for the case when longitudinal diffusion occurs, E. Wicke showed that the sorption front should be gradually smoothed out approximately in proportion to $\sqrt{t}$ (where t is time). In another work, E. Wicke [233] simplified the problem of the equilibrium dynamics of sorption still more and considered the case when longitudinal diffusion can be neglected. He established one of the most important quantitative principles of equilibrium dynamics of sorption, defining the dependence of the deformation of the front of sorption dynamics on the type of sorption isotherm.

J. Wilson [241] has shown that in the absence of effects of longitudinal transport in the chromatographic column, zones with sharp boundaries, without blurring of the fronts, should appear. He was the first to theoretically obtain a formula for the rate of motion of the stationary front of equilibrium sorption dynamics.

Ya. B. Zel'dovich's ideas were developed in the studies of O. M. Todes and associates [6-8, 116, 140, 142]. Further development of the theoretical investigations of E. Wicke and J. Wilson was contributed by De Vault [229] and by J. Weiss [211, 212, 232]. They made a detailed analysis of the influence of the type of sorption isotherm on the character of the motion of the concentration points of the front, character of the formation and deformation of the front of dynamic sorption. L. V. Radushkevich [105] showed that random packing of the grains in the layer of a real sorbent leads to a unique supplementary longitudinal transport of substances in the sorption column—the granulation and soil effects. An investigation of the problem of equilibrium sorption dynamics in the case of a linear isotherm, considering these effects, led L. V. Radushkevich to a solution that proved analogous to the solution obtained by E. Wicke [234, 235]. This indicated the formal possibility of considering the blurring of the sorption front under the action of granulation and soil effects by analogy with longitudinal diffusion, i.e., of considering them as quasidiffusion processes. Later, O. M. Todes and Ya. M. Bikson [6-8] proposed that a generalized coefficient be introduced to consider together all the kinetic and hydrodynamic factors of the blurring of the sorption front. The role of these factors varies under various conditions of the sorption process. But all the factors operate in the same direction—they create blurring of the fronts of the dynamically sorbed substances.

The formation in many cases of stationary sorption fronts and their parallel transport were well-known from experimental investigations in the field of sorption dynamics and chromatography. The first complete theoretical substantiation of the phenomenon of formation of a stationary front was given by Ya. B. Zel'dovich [63, 116, 139, 140]. He showed that a stationary front is formed only when the sorption isotherm is convex, while the degree of blurring of the stationary front depends on the kinetics of the sorption process.

A. A. Zhukhovitskii, Ya. L. Zabezhinskii, and A. N. Tikhonov [53, 61, 137] have developed a theory of the dynamics of nonequilibrium molecular sorption of gases and vapors. The equations of the linear isotherm and Langmuir isotherm were selected as the sorption isotherms, and the equation of diffusion kinetics of sorption as the equation of the sorption kinetics. For the initial stages of dynamic sorption, it proved impossible to obtain an exact analytical solution of the problem. Hence, the dynamic distribution curves for the stage of front formation were calculated by the method of finite differences. For the stage of parallel transport of the front, we obtained an asymptotic solution in analytical form.

These same authors gave a solution of the problem of nonequilibrium sorption dynamics for a linear isotherm, similar to the solution of the problem of mass exchange and heat transfer [17, 19, 74, 89, 132, 162, 172, 182, 218]. The same problem was solved in a somewhat different mathematical form in the studies of other authors [6, 153, 171, 202, 227]. The basic principle that follows from these solutions is the fact that in the case of nonequilibrium dynamics of sorption, for a linear isotherm, there should be a gradual blurring of the front of the substance to be sorbed. The blurring of the front is approximately proportional to the square root of the time ($\sim \sqrt{t}$).

J. Walter [230], H. Thomas [227], L. Sillen [180, 220-225], and V. L. Anokhin [3] developed a theory of nonequilibrium sorption dynamics for the case of chemical kinetics of sorption.

A variation of the approximate solution of the problems of sorption dynamics was proposed by S. E. Bresler and Ya. S. Uflyand [11-13]. They showed that sorption kinetics can be considered by the introduction of a special parameter—the lag time—into the equation of the sorption isotherm. The results obtained by S. E. Bresler and Ya. S. Uflyand in general agree with the previously developed theories of sorption dynamics for linear and convex isotherms.

L. A. Myasnikov and K. A. Gol'bert [97] attempted to solve the problem of nonequilibrium sorption dynamics, using the exact equation of intradiffusion kinetics as the equation of sorption kinetics and the equation of the linear sorption isotherm as the sorption isotherm. Operational calculus was used in the solution of the problem. However, the indicated authors could not obtain a solution in analytical form by this method for the problem posed. They had to use the method of numerical integration and represent the solution in the form of dimensionless curves.

L. V. Radushkevich [105] and N. N. Tunitskii et al. [146-150] showed that a number of important pieces of information on the dynamics of sorption (center of gravity of the sorption waves, average width, character of asymmetry of the sorption waves, degree of blurring, etc.) can be obtained using the methods of statistical physics. This trend in the theory of sorption dynamics and chromatography was developed further in the studies of Ya. V. Shevelev [163-165].

Close to the statistical method of describing sorption dynamics is the method of "theoretical plates," or the layer-by-layer method. The chromatographic column is considered as a set of some number of elementary finite layers of the sorbent, while the process of sorption dynamics itself is considered as an intermittent process of sorption from portions of the mobile phase, moving in batches from layer to layer. The method of theoretical plates was used by A. Martin and R. Synge [207] for the theoretical interpretation of partition chromatography.

In 1947-1948, E. N. Gapon and T. B. Gapon applied the layer-by-layer method to an approximate calculation of ion exchange chromatograms [31-33]. The indicated authors calculated the distribution of substances between the phases on the basis of the assumption of the establishment of instantaneous sorption equilibrium in the elementary layers. No theoretical estimate of the width of the elementary layer is given in these studies. The development of the mathematical apparatus of the layer-by-layer method and the solution of a number of problems of chromatography were the subjects of the works of V. V. Rachinskii [108-110] and V. P. Meleshko [94, 96].

We should emphasize the importance of the development of a theory for calculating chromatograms using the method of calculus of finite differences. In the solution of differential equations describing chromatographic processes, considerable mathematical difficulties are encountered. For example, it is impossible to obtain a solution in analytical form for the initial stages of the dynamics of the sorption process. The methods of calculus of finite differences are universal, permitting the solution of practical problems of sorption dynamics. The technical difficulties of the calculation can be overcome with the aid of electronic computing machines.

The question of estimating the width of the elementary layer or step, using the layer-by-layer method in calculations of dynamic distributions of substances in sorption columns, has been the subject of the works of Schubert, Mayer, and Tompkins [67, 92], N. N. Tunitskii [150], Ya. V. Shevelev [63], V. V. Rachinskii [115, 121], and V. P. Meleshko [94-96].

The theory of the sorption dynamics of mixtures of substances was first developed in our country by M. M. Dubinin and S. Yavich [46]. In the construction of the theory, these authors proceeded from the empirical principles established by N. A. Shilov [168].

O. M. Todes [139] gave a detailed theory of a steady-state system of nonequilibrium sorption dynamics of mixtures of substances. He showed that the velocities of individual sorption waves are arranged in an order of magnitude opposite to the relative sorbability of the components of the mixture. Moreover, each individual wave is propagated at a speed greater than that with which each individual substance would be propagated at the same initial concentration. The desorption action of the more strongly sorbed component of the mixture is manifested in this process. For the case of the sorption dynamics of two substances, the sorption of which obeys the Langmuir isotherm, quantitative principles of the process are given, formulas of the velocities of the fronts are derived, as are formulas for determining the concentrations of the components in the zones. For more complex systems, a solution is given in general form. S. Claesson and A. Tiselius [77, 228], independently of O. M. Todes [139], gave calculation formulas for the determination of the concentrations of substances in zones of a primary, frontal chromatogram, and on this basis developed an experimental method for the chromatographic analysis of mixtures of substances, which has received the title of frontal analysis.

J. Wilson [241] developed the bases of the theory of equilibrium chromatography. He has also given a solution for the problem of separating mixtures of two substances by the elution method. S. Claesson and A. Tiselius have contributed a theory of displacement chromatography for equilibrium sorption dynamics [77, 228].

Broad investigations of the theory of chromatography, developing the basic trends of the works of J. Wilson, D. De Vault, J. Weiss, and other authors, have been conducted by E. Glueckauf [186-197]. Ya. M. Bikson [8] was the first to calculate the concentration profiles of the primary chromatogram of two substances for nonequilibrium sorption dynamics. Questions of the theory of the desorption process were considered in the work of M. I. Yanovskii [169].

In recent years, the largest number of theoretical investigations has been devoted to the development of the theory of ion exchange chromatography [65, 94, 95, 112-114, 120-122, 128, 131, 142, 143, 153, 156, 186-197, 220-224], as well as gas chromatography [60, 76, 152]. This is explained by the fact that ion exchange and gas chromatography in recent years have acquired exceptionally broad practical application in varied branches of science and technology. There is a certain lag in the development of the theory of precipitation chromatography [99].

GENERAL FORMULATION OF THE PROBLEM OF THE DYNAMICS OF SORPTION AND METHODS OF ITS SOLUTION

1. Formulation of the Problem

The general theory of sorption dynamics and chromatography formulates the most general principles and methods, correct for any types of sorption and chromatography [117-119]. Let there be some medium with a highly developed surface, possessing sorption capacity, and let a flux of a mixture of substances be introduced into this medium. With the initial concentrations of the substances in the mixture, character of the interaction between the sorbent and components, as well as other conditions that influence the motion and distribution of the substances known, the problem consists of finding the space distribution functions of the substances in the sorbing medium for any moment of time. This is the direct problem of sorption dynamics. The opposite problem is also of great significance for chromatography: finding the initial composition of the mixture and concentrations of the substances to be analyzed in it by determining experimentally the distribution of the substances in the sorbing medium.

In the most general form, the theory of sorption dynamics should consider the following basic aspects of this complex physical phenomenon: the material balance during the motion and distribution of substances in the sorbing medium, the kinetics and statics of the sorption of substances, the hydrodynamics of the process, relationship between the thermodynamic parameters of the state of the medium, heat balance, and heat transfer during the process of sorption in a moving medium. The character of the motion and distribution of substances in the sorbing medium is also predetermined by the initial and boundary conditions of the process.

Two groups of methods, usually used for the theoretical solution of physical and chemical problems, are used for the quantitative, mathematical description of the process of sorption dynamics—the phenomenological and statistical methods. Phenomenological methods establish the functional relationships among the quantities characterizing the physical phenomenon from the macroscopic point of view. In a phenomenological description of the process, no effort is made to give a molecular-kinetic, microscopic interpretation of the phenomenon and the quantities that characterize it. The methods of statistical physics, in contrast to phenomenological methods, give a molecular-kinetic, statistical interpretation of the physical phenomenon, characterizing its quantities and the relationships among them. In the phenomenological approach to the solution of the problems of sorption dynamics, one proceeds from the basic assumption that the process of sorption dynamics is a continuous process. The sorbent is considered as some liquid- or gas-permeable phase, in which the "sorption activity" of the sorbent is continuously and uniformly distributed.

In search of the distribution functions of the substances in the sorbing medium sought at any stage of the process, one compiles a system of differential equations, quantitatively describing the process of dynamic sorption. As will be seen later, systems of differential equations describing the processes of sorption dynamics in the most general form are differential equations in partial derivatives of the first and second orders. The solution of such systems of equations at set boundary conditions is the subject of differential and integral calculus, in particular, the disciplines that have received the name of mathematical physics [15, 80, 91, 133].

The systems of differential equations of sorption dynamics are solved with great difficulty in the final analytical form. Only in particular cases, and, moreover, with simplifying assumptions, can full analytical solutions be obtained. In those cases when the methods of mathematical physics do not permit the solution of problems of sorption dynamics, one has recourse to a numerical integration of the differential equations, i.e., the methods of the calculus of finite differences are used for the solution [100, 101].

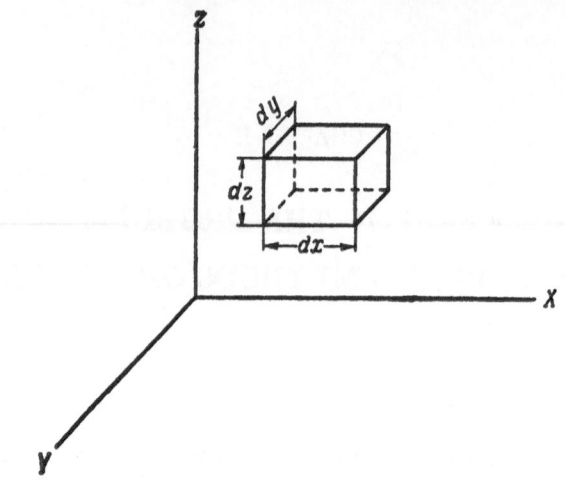

Fig. 1. For derivation of the equation of the material balance.

In the statistical solution of the problems of sorption dynamics, the sorbent is considered as a discrete medium with randomly arranged sorption centers, through which a statistical assemblage of sorbed particles moves in a flux. The field of velocities of the particles, in spite of the general directedness of the flux, is random in character. The sorption process is characterized by probability functions for the location of each particle in the phase of sorbent or the mobile phase. These probability functions also serve as a basis for calculating the statistical distribution of the particles in the sorbing medium.

2. Equations of the Material Balance

Let us consider the compilation of the equations of the material balance during their motion through a porous medium. Let a flux of a mixture of substances to be sorbed be moving in any direction within a porous medium.

We shall assume that the average porosity and sorption properties of the medium are the same at all points.

Let us introduce the following notation:

X, Y, Z — rectangular coordinates, in which the motion of the substances is considered;

t — time of the process;

u — linear velocity of the flux within a porous medium;

$\vec{u}$ — vector quantity corresponding to u;

j — number of components of the mixture;

i — ordinal number of the component of the mixture;

n_i — volume concentration of the i-th component in the composition of the mobile phase, calculated per unit volume of the porous medium;

N_i — volume concentration of the i-th component in the stationary phase (sorbent), also calculated per unit volume of the porous medium.

The flux velocity $\vec{u}$ and concentrations n_i and N_i are functions of the coordinates and time.

Let us isolate an elementary infinitesimal volume dxdydz in the porous sorbing medium (volume of an infinitesimal parallelepiped), as shown in Fig. 1.

Let us consider the material balance of the i-th component. Let us break up the material flux into three components of the flux, characterized by the corresponding components of the vector of the linear velocity: u_x, u_y, and u_z. The amount of matter entering the elementary volume through the face dydz in a time dt will be equal to $u_x n_i dydzdt$, where n_i is the concentration of the i-th component at the entrance to the elementary volume. The amount of matter exiting through the opposite face of the parallelepiped will be equal to

$$u_x n_i dydzdt + \partial/\partial x \, (u_x n_i \, dydzdt)dx.$$

The change in the amount of matter in the elementary layer during a time dt as a result of the passage of the material flux in the direction of the X-axis will be equal to the difference between the amount of matter that entered the elementary volume and that which left it, i.e.,

$$- \frac{\partial}{\partial x} (u_x n_i dydzdt)dx = - \frac{\partial}{\partial x} (u_x n_i)dxdydzdt.$$

The change in the amount of matter in the direction of the Y-axis is considered analogously: $-(\partial/\partial y)(u_y n_i)dxdydzdt$, and in the direction of the Z-axis: $-(\partial/\partial z)(u_z n_i)dxdydzdt$.

Summing the changes in the amount of matter in the elementary layer in all three directions, we obtain a total change equal to

$$- \left[\frac{\partial (u_x n_i)}{\partial x} + \frac{\partial (u_y n_i)}{\partial y} + \frac{\partial (u_z n_i)}{\partial z} \right] dxdydzdt = - [\mathrm{div}\ (n_i \vec{u})]dxdydzdt.$$

The change obtained in the amount of matter in the elementary layer gives rise to the corresponding changes in the amount of matter in the sorbent and mobile phase. The change in the amount of matter in the sorbent will be equal to $(\partial N_i/\partial t)dtdxdydz$. Correspondingly, the change in the amount of matter in the mobile phase will be equal to $(\partial n_i/\partial t)dtdxdydz$.

Now we can lead up to the total material balance in the elementary layer:

$$- [\mathrm{div}\ (n_i \vec{u})]dxdydzdt = \frac{\partial N_i}{\partial t} dxdydzdt + \frac{\partial n_i}{\partial t} dxdydzdt$$

or

$$\frac{\partial n_i}{\partial t} + \frac{\partial N_i}{\partial t} + \mathrm{div}\ (n_i \vec{u}) = 0. \qquad (\text{II.1})$$

If instead of a mechanical flux, there is a supplementary transport of matter as a result of molecular diffusion and convection, then another supplementary factor which considering these processes of matter transport must be introduced into equation (II.1).

Let us consider diffusion-convection transport according to the theory of diffusion [82] by the following equation:

$$\left(\frac{\partial n_i}{\partial t} \right)_{\mathrm{diff}} = D_{m,i} (\Delta n_i + \lambda_i \Delta T), \qquad (\text{II.2})$$

where $D_{m,i}$ is the coefficient of molecular diffusion; λ_i is the coefficient of thermal diffusion; Δ is the Laplace operator.

The quantity $(\partial n_i/\partial t)_{\mathrm{diff}} dxdydzdt$ characterizes the supplementary change in the amount of matter in the elementary volume on account of the phenomenon of molecular diffusion and convection. Then the total equation of the material balance in the process of sorption dynamics will take the following form:

$$\frac{\partial n_i}{\partial t} + \frac{\partial N_i}{\partial t} + \mathrm{div}\ (n_i \vec{u}) = D_{m,i} (\Delta n_i + \lambda_i \Delta T). \qquad (\text{II.3})$$

For the process of sorption dynamics of a mixture, we shall have a system j of such equations of balance $(1 \le i \le j)$.

3. Equations of the Kinetics and Statics of Sorption

The following equations, characterizing the dynamics of sorption, should be equations of the kinetics of sorption, reflecting the physicochemical nature of the sorption process and establishing a temporary relationship between the concentrations of the substances in the sorbent and in the mobile phase [136].

In the most general case, these should be complex integral-differential equations (functionals), since the concentrations of the components to be sorbed at a given moment of time can be functions of the concentrations of the components not only at the given moment, but also in the preceding moments of time. Such complex equations, considering all aspects of the kinetics of the sorption equation, are rather unsuitable for practice. Hence, approximate equations must be compiled, considering the main factors of sorption kinetics.

Proceeding from the most general considerations, we can assume that the kinetics of the sorption of any i-th component depends on the following basic independent factors: on the concentrations of all the components, both in the mobile phase and in the sorbent ($n_1, n_2, \ldots, n_j; N_1, N_2, \ldots, N_j$); on the parameters determining the so-called diffusion step of sorption kinetics (among them let us mention the influence of the rates of flow of the mobile phase $\vec{u}$, temperature T, and density of the mobile phase ρ; we shall symbolically denote the remaining parameters as K_{diff}); of the parameters determining the so-called chemical step of the sorption kinetics (direct interaction of the sorbent and particles to be sorbed), we shall symbolically denote the indicated parameters as K_{chem}.

Considering the factors noted, let us write the equation of the sorption kinetics in the form

$$\frac{\partial N_i}{\partial t} = \Psi_i \, (n_1, \ n_2, \ldots, \ n_j; \ N_1, \ N_2, \ldots, \ N_j; \ \vec{u}, \ T, \ \rho, \ K_{diff} \, K_{chem}). \tag{II.4}$$

In this equation, to the left we have the partial derivative of the concentration of the i-th component in the sorbent with respect to time, since in the case of sorption dynamics, N_i is a function not only of the time, but also of the coordinates (X, Y, Z). For static conditions of sorption (absence of directed motion of substances with respect to the sorbent), the concentration N_i is a function only of the time, and then the usual derivative dN_i/dt should be written in equation (II.4).

When the concentrations $N_1, N_2, \ldots, N_j$ depend not only on the concentrations $n_1, n_2, \ldots, n_j$ at a given moment, but also on those at the preceding moments, in the right-hand portion of equation (II.4), Ψ_i is denoted by a functional. If the concentrations $N_1, N_2, \ldots, N_j$ depend only on the concentrations $n_1, n_2, \ldots, n_j$ at a given moment, then Ψ_i is denoted by the usual function.

As a rule, the sorption process is reversible, and sorption is accompanied by desorption. The resultant rate of sorption dN_i/dt is equal to the difference between the rates of the forward process (sorption itself) and the reverse process (desorption)

$$\frac{\partial N_i}{\partial t} = \left(\frac{\partial N_i}{\partial t}\right)_{forw} - \left(\frac{\partial N_i}{\partial t}\right)_{rev} .$$

When all the physical factors upon which the sorption depends (u, t, ρ, K_{diff}, K_{chem}) remain constant during the sorption process, the rate of the sorption process will depend only on the concentrations of the substances to be sorbed:

$$\frac{\partial N_i}{\partial t} = \Psi_i \, (n_1, \ n_2, \ldots, \ n_j; \ N_1, \ N_2, \ldots, \ N_j). \tag{II.4'}$$

For heterogeneous processes, two limiting cases of heterogeneous interactions are distinguished: diffusion kinetics and chemical kinetics [83]. The same approach can also be applied to the kinetics of sorption processes.

In the case of diffusion kinetics of sorption, the rate of sorption is limited by the rate of diffusion of the particles to be sorbed to the sites of sorption. In this case, the rate of sorption will depend chiefly on the rate of flow, density, viscosity, mass of the particles, and geometrical factors determining the rate of diffusion. The diffusion kinetics of sorption in the case of intermicellary sorption (the grains of the sorbent possess internal porosity) is subdivided into two stages: external and internal diffusion.

At the external diffusion stage, the stage of diffusion of the particles to be sorbed into the mobile phase, the hydrodynamic system under which the sorption process is carried out exerts a vital significance. The rate of internal diffusion does not depend on the hydrodynamic system, but it is influenced by a number of parameters, especially geometrical, characterizing molecular diffusion within the pores of the sorbent.

In the chemical kinetics of sorption, the deciding influence on the rate of the process is exerted by the nature of the forces of interaction between the sorbent and the particles to be sorbed, while the mechanical properties of the medium and geometrical factors become relatively unimportant.

The concrete forms of the kinetic equations will not be considered here. In Section 7 we shall consider approximate forms of the equations.

As has already been noted above, in the phenomenological approach, the molecular-kinetic, statistical picture of the process is not considered in the description of sorption processes. Although the equations of sorption kinetics reflect the physicochemical nature of the sorption process, however, the phenomenological equations of diffusion mass transfer also represent the kinetics of chemical reactions.

Both under static and under dynamic conditions, a sorption equilibrium may be established in the process of reversible sorption, in which the concentrations of all the components in the system become constant. The relationship between the equilibrium concentrations of the substances to be sorbed in the composition of the sorbent and mobile phase is determined by the equation of the sorption statics or by the equations of the sorption isotherms at constant temperature ($T = const$). In a number of cases, the equations of the sorption isotherms can be obtained from the equations of the sorption kinetics. For example, if the equation of the sorption kinetics describes a sorption process in an isolated sorbent-solution system (static conditions), then in time such a system should come to an equilibrium state. The rate of sorption dN_j / dt for the state of equilibrium will be equal to zero. This means that the rate of the forward and reverse processes—sorption and desorption—are equal at equilibrium. Then function ψ_i is set equal to zero in the equation of the sorption kinetics (II.4), while the concentrations of the components will be equilibrium. The equation of the sorption statics can be obtained in this way from the equation of the sorption kinetics.

A characteristic feature of the equations of the sorption isotherms is the fact that they do not contain any parameters upon which the sorption kinetics depends, for example, the parameters determining diffusion. The equations of the sorption isotherms contain certain constants, the so-called constants of sorption statics, characterizing the sorbability of the substances.

In general, the equations of the sorption isotherms can be written in the form of the following implicit functions:

$$F_i(n_1, n_2, \ldots, n_j; N_1, N_2, \ldots, N_j) = 0. \tag{II.5}$$

In certain cases, the sorption isotherms can be obtained in the form of explicit functions:

$$N_i = f_i(n_1, n_2, \ldots, n_j). \tag{II.6}$$

In the solution of the problems of sorption dynamics and chromatography, to simplify the problem (and in many cases this simplification is close to the real conditions), one neglects the kinetics of sorption, assuming that sorption equilibrium is established practically instantaneously. Then, the equations of the sorption isotherms will be contained in the system of differential equations of sorption dynamics in place of the equations of the sorption kinetics.

Hence, two cases of sorption dynamics are possible: equilibrium and nonequilibrium.

4. Equations of Hydrodynamics

The value of the rate of flow of the mobile phase is contained in the equation of the material balance of the substances to be sorbed (II.3), and in general in the kinetic equation (II.4) as well. This means that the distribution of the substances to be sorbed during the process of sorption dynamics depends on the distribution of velocities in the porous medium in space and time, i.e., $\vec{u}(x, y, z, t)$. The space-time distribution functions of the rate of flow can be sought on the basis of the conditions of hydrodynamics [81-83, 133].

Let us recall somewhat briefly some of the most general information from hydrodynamics.

The motion of a liquid or gas can be described mathematically with the aid of mutually dependent functions of the rate $\vec{u}(x, y, z, t)$, density $\rho(x, y, z, t)$, pressure $p(x, y, z, t)$, and temperature $T(x, y, z, t)$.

From the macroscopic point of view, a liquid or gas is considered in hydrodynamics as a continuous medium. Of all three indicated quantities—velocity, density, and pressure—only the velocity is a vector quantity. The density, pressure, and temperature are scalars. From this it follows that six quantities will be unknown: the three components of the velocity—u_x, u_y, and u_z, and three scalars—the density ρ, pressure p, and temperature T. Hence, we need a system of six equations to search for the unknown functions.

The first equation of this system will be the equation of discontinuity:

$$\frac{\partial \rho}{\partial t} + \text{div}\,(\rho \vec{u}) = 0. \tag{II.7}$$

This is a scalar equation—the expression of the law of conservation (balance) of the mass of the fluid phase (liquid or gas).

The following equation is the equation of motion of a viscous liquid:

$$\frac{\partial \vec{u}}{\partial t} + (\vec{u}\nabla)\,\vec{u} = -\frac{\nabla p}{\rho} + \frac{\vec{F}}{\rho} + \frac{\mu}{\rho}\,\Delta \vec{u} + \frac{1}{\rho}\left(\zeta + \frac{\mu}{3}\right)\nabla\,\text{div}\vec{u}. \tag{II.8}$$

This is a vector equation—the expression of the second law of Newtonian dynamics: the left side represents the acceleration $d\vec{u}/dt$ of a unit mass, while the right side represent the forces acting on this mass. The factor $\nabla p/\rho$ defines the value of the hydrostatic pressure acting on the fluid phase; $\vec{F}$ is the external force; the remaining factors characterize the force of internal friction (ζ and μ represent the coefficients of viscosity of the medium, depending on the temperature of the medium T). The vector equation (II.8) breaks down into three equations when projected on the coordinate axes X, Y, Z.

The fifth in the system of hydrodynamic equations is the equation of state of the liquid (or gas):

$$\rho = f\,(p,\ T), \tag{II.9}$$

which establishes the dependence of the density of the mobile phase on the pressure and temperature for a given state of aggregation of the medium.

During the process of sorption dynamics, the mobile, flowing phase represents liquid solutions or mixtures of gases. The composition of the liquid (or gaseous) phase will vary continuously in this process, and, consequently, the density of the mobile phase ρ will be not only a function of the pressure and temperature, but also a function of the concentrations of the components of the mobile phase:

$$\rho = f\,(n_1, \ldots, n_j;\ p, T). \tag{II.10}$$

The sixth equation of hydrodynamics is the equation of the balance and propagation of heat.

The field of velocities and temperatures is the result of a combination of mechanical and thermal interactions; strictly speaking, they are interdependent. The temperature field influences the velocity field, both through the dependence of the density ρ on the temperature, and through the temperature dependence of the viscosity. On the other hand, the temperature field depends on the velocity field, since when the fluid phase is transported, a transfer of heat is also accomplished simultaneously.

The equation of heat propagation in a moving medium takes the general form

$$\text{div}\,(\lambda\nabla T) + q + A\mu\ \text{diss}\,F\,(\vec{u}) = c\rho\,\frac{DT}{dt} - A\,\frac{Dp}{dt}, \tag{II.11}$$

where λ is the coefficient of thermal conductivity; c is the specific heat in a set thermodynamic process; q is the density of the internal source (or outflux) of heat; A is the heat equivalent of mechanical work.

The equation of the balance and transfer of heat (II.11) is solved together with the system of equations of hydrodynamics (II.7)-(II.9). The internal sources or absorbers of heat q may be exothermic or endothermic sorption processes, accompanied by the evolution or absorption of heat. Under the conditions of motion of a viscous liquid, dissipation of mechanical energy occurs, along with the liberation of heat at its expense, as a result of the phenomenon of internal friction.

The influence of temperature is not limited only to its interaction with the velocity field. The transport of matter during the process of sorption dynamics also depends on the temperature field. In general, the equation of balance (II.3) contains the factor of thermal diffusion transport of the substances to be sorbed, which depends on the temperature gradient. The temperature also influences the kinetics and statics of sorption.

5. Initial and Boundary Conditions

For a complete formulation of the problem of sorption dynamics, one must set not only the differential equations, but also the initial and boundary conditions. The initial conditions characterize the distribution functions of the concentrations of the substances to be sorbed, densities, velocities, and temperatures in the system of sorbent and mobile phase at the initial moment of time (t = 0). The boundary conditions define these functions at the boundaries of the system as a whole, as well as on the boundaries between phases. Depending on the concrete physical conditions in which the system is found, the initial and boundary conditions can be most varied. The development of the process of sorption dynamics depends unambiguously on the initial and boundary conditions. Hence, the proper formulation of the initial and boundary conditions is an inseparable portion of the mathematical formulation of the problem of sorption dynamics.

Let us mention that from the practical standpoint, the initial and boundary conditions reflect different methods of conducting the process of sorption dynamics and chromatography. For example, frontal chromatography is characterized by the fact that constant initial concentrations of the substances to be sorbed are always maintained at the entrance to the chromatographic column. In elution chromatography, some sort of solvent is introduced at the entrance to the column containing the sorbed substances, and, consequently, the concentration of the components in the mobile phase at the entrance to the column is kept equal to zero.

In displacement chromatography, a solution of the component—displacer—is introduced at the entrance to the column containing the sorbed substances, and, consequently, a constant concentration of the displacer is created at the entrance to the column in the mobile phase. In the case of gradient elution or displacement, a definite variation of the concentration of the eluting solution with the course of time is created at the entrance to the column.

In the consideration of the concrete problems of sorption dynamics, various initial and boundary conditions will be formulated mathematically and considered in great detail.

6. Simplification of the System of Equations of Sorption Dynamics

The system of equations of the material balance of substances to be sorbed (II.3), kinetics or statics of sorption (II.4)-(II.6), and the equations of hydrodynamics (II.7)-(II.11), together with the initial and boundary conditions, describe in most general form the process of sorption dynamics.

The solution of this system of equations at set initial and boundary conditions in principle should give the functions of the space-time distribution of the sorbed substances in the sorbing medium, of interest to us. These distribution functions sought may be explicit or implicit:

$$F_{n,i}\,(n_i,\ x,\ y,\ z,\ t) = 0; \tag{II.12}$$

$$F_{N,i}\,(N_i,\ x,\ y,\ z,\ t) = 0 \tag{II.13}$$

or

$$n_i = f_{n,i}\,(x,\ y,\ z,\ t); \tag{II.14}$$

$$N_i = f_{N,i}\,(x,\ y,\ z,\ t). \tag{II.15}$$

19

At the present time, it is impossible to give a general solution for the equations of sorption dynamics, and the theory thus far has been developed along the line of simplifying the equations, introducing various assumptions and approximations, and considering the simplest particular cases. Let us consider the conditions and assumptions under which various simplifications of the equations of sorption dynamics can be made.

1. Let us assume that the process of sorption dynamics is isothermal [$T(x, y, z, t) = $ const]; consequently, there are not temperature gradients ($\Delta T = 0$). In this case the process of thermal diffusion will be absent, and the equation of balance of the substances to be sorbed takes the form

$$\frac{\partial n_i}{\partial t} + \frac{\partial N_i}{\partial t} + \text{div}\,(n_i \vec{u}) = D_{m,i}\,\Delta n_i. \tag{II.16}$$

The temperature argument in the equation of the sorption kinetics (II.4) and in the equation of state (II.9) drops out. An isothermal character of the process of sorption dynamics can be created by selection of all the processes of outflux, influx, evolution and absorption of heat, such that a constant temperature T is established at all points of the system. Thus, the conditions $T(x, y, z, t) = $ const actually should be the solution of the system of equations of hydrodynamics with respect to the temperature in the case of definite boundary conditions and properties of the system, guaranteeing rapid equilibration of the temperature within the system. In the case of an isothermal process, equation (II.11) drops out; it is replaced by the condition $T(x, y, z, t) = $ const.

Let us mention that the isothermal character of the process of sorption dynamics under real conditions can be achieved only approximately, in view of the finite character of the rates of transfer, evolution, and absorption of heat. The condition or solution $T(x, y, z, t) = $ const thus is approximate and asymptotic in character.

The questions of the transfer of heat in porous media and sorption columns have been discussed in detail in a number of studies [19-22, 89, 144, 162].

2. Let us assume that the mobile phase is incompressible, and the concentrations of the substances to be sorbed are so small that the variation of the density as a function of the concentration of the substances can be neglected ($\rho = $ const).

Under this condition, the equation of continuity (II.7) takes the form

$$\text{div}\,\vec{u} = 0. \tag{II.17}$$

Considering that $\rho = $ const, let us use equation (II.17) to transform equation (II.16):

$$\frac{\partial n_i}{\partial t} + \frac{\partial N_i}{\partial t} + \vec{u}\nabla n_i = D_{m,\,i}\,\Delta n_i. \tag{II.18}$$

When $\rho = $ const, the equation of motion of a viscous liquid (II.8) is also simplified:

$$\frac{\partial \vec{u}}{\partial t} + (\vec{u}\nabla)\,\vec{u} = -\frac{\nabla p}{\rho} + \frac{\vec{F}}{\rho} + \frac{\mu}{\rho}\,\Delta\vec{u}. \tag{II.19}$$

The latter equation is called the Navier-Stokes equation.

3. The equation of balance of the substances to be sorbed (II.18) and the equations of hydrodynamics (II.17)-(II.19) cannot be used in practice in the form in which they were written to solve the problem of the dynamics of sorption of substances in porous media. Actually, very complex velocity fields arise within the channels of the porous medium. In the presence of statistical inhomogeneity of the pores of the sorbing medium, the distribution of the rates of flow will also be statistical in character (granulation effect). Hence, it is advisable to introduce some average velocity $\vec{u}$, while the probability of deviation of the real velocities from this average velocity can be considered by the introduction of the quasidiffusion coefficient $D_{l,i}$, which characterizes the supplementary longitudinal transport of matter along the line of flow. In this case, the equation of the balance (II.18) takes the form

$$\frac{\partial n_i}{\partial t} + \frac{\partial N_i}{\partial t} + (\vec{u},\ \nabla n_i) = D_i^*\,\Delta n_i, \tag{II.20}$$

where $D_i^* = D_{m,i} + D_{l,i}$ is the effective quasidiffusion coefficient of longitudinal transport, considering all the effects of statistical longitudinal transport; $\vec{u}$ is the average rate of transport at a set point at a given moment. Thus, the introduction of the average velocity reduces the problem of the motion of a mobile viscous phase in a porous medium to the usual hydrodynamic problem of the motion of a viscous continuous medium. In this case, of course, some effective force of friction, considering the hydraulic resistance of the porous medium [84], must be introduced into the equation (II.8).

4. For further simplification of the equations of sorption dynamics, let us assume that the motion of the flux is accomplished only in one direction (unidimensional problem), for example, in the direction of the axis OX, with an average constant velocity $u = const$. In this case, the problem is considerably simplified: only the equations of the material balance and the equations of sorption kinetics remain in the system of equations of sorption dynamics. The equation of the balance for the unidimensional problem will take the form

$$\frac{\partial n_i}{\partial t} + \frac{\partial N_i}{\partial t} + u\,\frac{\partial n_i}{\partial x} = D_i^*\frac{\partial^2 n_i}{\partial x^2}\,. \tag{II.21}$$

The average linear velocity of the flux of the mobile phase can be determined experimentally according to the formula

$$u = \frac{V}{Qt}, \tag{II.22}$$

where V is the volume of the mobile phase, introduced into a porous medium in a time t; Q is the average cross-sectional area of transport of the substance to be sorbed within the porous medium.

5. As has already been noted above, the compilation and substantiation of exact equations of sorption kinetics is a complex physical problem. Hence, in the theory of sorption dynamics and chromatography, one takes the line of compiling simplified approximate phenomenological equations of sorption kinetics. Let us consider a series of these methods of compiling approximate equations.

First let us take the two extreme cases of sorption kinetics: chemical and diffusion kinetics. In general, the equation of chemical sorption kinetics can be represented in the following form:

$$\frac{\partial N_i}{\partial t} = k_1\varphi_1\,(n_1,\,n_2,\,\ldots,\,n_j;\ N_1,\,N_2,\,\ldots,\,N_j) - k_2\varphi_2\,(n_1,\,n_2,\,\ldots,\,n_j;\ N_1,\,N_2,\,\ldots,\,N_j), \tag{II.23}$$

where k_1 and k_2 are the constants of the sorption kinetics of the forward and reverse reactions, respectively; φ_1 and φ_2 are functions of the concentrations of the components, depending on the mechanism of the heterogeneous interaction. The following approximate equation is usually used to describe the diffusion kinetics:

$$\frac{\partial N_i}{\partial t} = \beta_i\,(n_i - n_i'), \tag{II.24}$$

where β_i is the effective rate constant of diffusion; n_i' is functionally related to N_i by the sorption isotherm. The rate constant of diffusion considers the external and internal diffusion in sum (in the case of intramicellary sorption), so that

$$\frac{1}{\beta_i} = \frac{1}{\beta_i'} + \frac{1}{\beta_i''}\,, \tag{II.25}$$

where β' is the rate constant of external diffusion; β'' is the rate constant of internal diffusion.

The physical meaning of equation (II.24) consists of the fact that the motive force of diffusion sorption kinetics is found in the concentration gradients between points far from the site of sorption and points of the interface where the events of sorption are carried out. This means that the nonequilibrium concentration n_i far from the interface should be greater than the concentration at the interface. In equation (II.24), it is assumed that the concentration of the component at the interface n_i' is an equilibrium concentration with respect to the concentration of this component in the sorbent N_i.

6. The simplest case of sorption dynamics is the dynamics of the sorption of one component. The system of equations of the sorption dynamics of one substance, under the simplifying conditions assumed above, will take the form

$$\frac{\partial n}{\partial t} + \frac{\partial N}{\partial t} + u\,\frac{\partial n}{\partial x} = D^* \,\frac{\partial^2 n}{\partial x^2};$$

(II.26)

$$\frac{\partial N}{\partial t} = \Psi\,(n,\ N).$$

(II.27)

Only the variables remain in the equation of sorption kinetics (II.27). Moreover, it is assumed that all other physical factors upon which the sorption depends remain constant during the process of sorption dynamics. In an equilibrium system of sorption dynamics, the kinetic equation should be replaced by the equation of the sorption isotherm, which can be written in the form of an implicit function:

$$F(n,\ N) = 0$$

(II.28)

or in the form of an explicit function:

$$N = f(n).$$

(II.29)

7. In concluding our discussion of various simplifications of the equations of sorption dynamics, let us consider still another method of compiling an approximate equation of sorption kinetics—the method of "lagging coordinates."

In this method, to avoid most of the varied physical constants determining the sorption kinetics, one has recourse to the use of a modification of the equation of the sorption isotherm by a certain displacement in the variables of the coordinates of the path x or time t [12, 163].

For simplicity, let us take the equation of the sorption isotherm of one component in the explicit form (II.29). In the method under consideration, the nonequilibrium concentration N(x, t) can be defined as the equilibrium concentration for a point displaced a distance $\varkappa$ along the X-axis ("lag path"):

$$N\,(x,\ t) = f\,[n\,(x + \varkappa,\ t)].$$

(II.30)

The sorption kinetics can also be considered by another method. The nonequilibrium concentration N(x, t) can be defined as the equilibrium concentration for a moment of time differing from the given moment of time by the quantity τ ("lag time"):

$$N\,(x,\ t) = f\,[n\,(x,\ t - \tau)].$$

(II.31)

The constants $\varkappa$ and τ introduced are certain phenomenological constants, together considering the influence of various conditions on the sorption kinetics, i.e., they possess the physical meaning of some phenomenological constants of sorption kinetics.

If we assume that the kinetic constants $\varkappa$ and τ are sufficiently small in comparison with the distance x traversed by the substance, or, correspondingly, with the elapsed time of sorption dynamics t, then equations (II.30) and (II.31) can be resolved in a series according to the parameters $\varkappa$ and τ, respectively. Cutting off these series at the second members and differentiating the functions obtained with respect to time, we can obtain the following approximate equations of the sorption kinetics:

$$\frac{\partial N}{\partial t} = \frac{\partial f}{\partial t} + \varkappa\,\frac{\partial^2 f}{\partial x \partial t};$$

(II.32)

$$\frac{\partial N}{\partial t} = \frac{\partial f}{\partial t} - \tau \frac{\partial^2 f}{\partial t^2} \cdot \qquad (\text{II.33})$$

7. Methods of Solving the Problems of Sorption Dynamics

The mathematical basis of the solution of the equations of sorption dynamics is the theory of equations in partial derivatives. This is one of the most difficult and still comparatively little developed divisions of higher mathematics. In the solution even of simplified equations of sorption dynamics, great mathematical difficulties are encountered. As a rule, only asymptotic solutions are obtained with comparative ease. The full solutions, describing all stages of the process of sorption dynamics, frequently cannot be found by the methods of analytical integration.

Various methods of the theory of differential equations in partial derivatives can be used to solve the equations of sorption dynamics. For example, the method of selecting new variables and using them to reduce the equations of sorption dynamics to a form for which the solution is already known is used. In a number of cases, the system of equations in partial derivatives can be reduced to equations in normal derivatives by replacement of variables.

One of the most widespread methods of solving equations in partial derivatives is the method of characteristics. In certain cases, analytical solutions can be obtained by the methods of operational calculus [97]. When analytical solutions of the equations of sorption dynamics cannot be obtained, one resorts to numerical methods of their solution [31-33, 53, 61, 94-95, 110, 121]. In view of this, a change-over is already being made to the solution of the equations of sorption dynamics with electronic computers. Our task does not include a consideration of the mathematical bases of the solution of equations in partial derivatives, which include the equations of sorption dynamics. For a detailed study of the theory of equations in partial derivatives, the reader must turn to the special literature [15, 80, 91, 101, 102, 133]. In the exposition of the concrete problems of the theory of sorption dynamics and chromatography, we shall use various methods of solving the equations of sorption dynamics.

In physics and technology, in those cases when exact analytical solutions of the physical and technological problems that arise cannot be obtained, the methods of the theory of similarity and model study are widely used [74, 81-84, 141]. Let us take up a number of general problems of the use of the method of similarity in the theory of sorption dynamics and chromatography.

First of all, let us mention that in the solution of physical problems, derivation or compilation of physical formulas and equations, there is always the possibility of imparting the greatest generality to the formulas and equations. One of the means of achieving such generality is the method of reducing these formulas and equations to dimensionless quantities. The formulas and equations in dimensionless quantities obtained in this case are called reduced.

The transformation of equations to the reduced form is accomplished by selecting the corresponding new scales for the dimensional quantities and the measurement of these quantities in the scales selected. The introduction of new scales for the measurement of dimensional quantities leads to the appearance of these selected scales in the equation of the complex. By suitable algebraic transformations, dimensionless coefficients consisting of the complex of selected scales and other constants of the equations can be obtained in these equations. These dimensionless coefficients have received the name of similarity criteria or criterial coefficients.

As an example, let us consider the system of equations of the sorption dynamics of one substance for the case of diffusion sorption kinetics, namely—for the transformation, let us say, of equations (II.26) and (II.24). Let us introduce the following characteristic scales for the variables contained in the equation: the characteristic concentrations n_0 and N_0, the characteristic linear dimension x_0, the characteristic time t_0, and the characteristic rate u_0. After the introduction of these new scales, we obtain the following reduced equation of the material balance:

$$\frac{\partial \varphi}{\partial \tau} + H_0 U \frac{\partial \varphi}{\partial X} + \frac{1}{h} \frac{\partial \vartheta}{\partial \tau} = F_0 \frac{\partial^2 \varphi}{\partial X^2} \,, \qquad (\text{II.34})$$

where $\varphi = n/n_0$, $\vartheta = N/N_0$, $U = u/u_0$, $X = x/x_0$, $\tau = t/t_0$, $H_0 = u_0 t_0/x_0$ is the criterion of homochrony; $h = n_0/N_0$ is the criterion of the distribution (distribution ratio); $F_0 = D^* t_0/x_0^2$ is the Fourier criterion.

The equation of the sorption kinetics (II.24) can be transformed analogously:

$$\frac{\partial \vartheta}{\partial \tau} = h \, Kn \, (\varphi - \varphi'), \tag{II.35}$$

where $Kn = \beta t_0$ is the criterion of the sorption kinetics.

The number of criteria in equations (II.34) and (II.35) can be reduced by the selection of suitable scales. For example, by introducing the following scales: $x_0 = u/\beta$ and $t_0 = 1/\beta$. Then we obtain

$$H_0 \cdot U = 1, \quad Kn = 1, \quad F_0 = \frac{D^* \beta}{u^2}.$$

The system (II.34), (II.35) in this case takes the form of the following dimensionless reduced equations:

$$\frac{\partial \varphi}{\partial \tau} + \frac{\partial \varphi}{\partial X} + \frac{1}{h} \frac{\partial \vartheta}{\partial \tau} = F_0 \frac{\partial^2 \varphi}{\partial X^2}, \tag{II.36}$$

$$\frac{\partial \vartheta}{\partial \tau} = h \, (\varphi - \varphi'). \tag{II.37}$$

According to the theory of similarity, two physical processes described by a given equation or system of equations are similar if the criteria of similarity in the reduced equations are equal, with identical initial and boundary conditions, as well as similarity of the geometrical parameters of the system.

The dimensionless criteria of similarity, present in the equations in the form of coefficients in front of the variables, are called determining in the theory of similarity. Other dimensionless independent variables — arguments — are also determining parameters. The dimensionless dependent variables of the reduced equation are functions of the determining criteria and arguments; they are called determined.

Thus, for example, for the system of equations (II.36), (II.37), the determining criteria and arguments are h, F_0, X, and τ, while the determined quantities are φ and ϑ:

$$\begin{aligned} \varphi &= f_\varphi \, (h, \; F_0, \; X, \; \tau); \\ \vartheta &= f_\vartheta \, (h, \; F_0, \; X, \; \tau). \end{aligned} \tag{II.38}$$

Of course, the algebraic transformation of the equations and its reduction to dimensionless form still is not a solution of the equations. Such a transformation imparts only greater generality, universality to the equation, since all the quantities that it contains are dimensionless. But it is just as difficult to obtain a solution for the equations of sorption dynamics in dimensionless form as for the original equations with dimensional quantities. The dimensionless equations form the theoretical base for the methods of experimental study of the processes.

The method of model study consists of an experimental accomplishment of a physical process that is similar to the process to be studied, and, on the strength of the similarity, a description by the same system of equations as for the process studied. More concisely, a model experiment, which artificially reproduces the real physical process, is conducted. In particular, as such a model experiment one can carry out the process on a small-scale laboratory setup, similar in all the basic parameters to the large setup planned. The values of the determining criteria of similarity are obtained with the aid of model experiments, and then, on the basis of the principle of similarity, using the values of the determining criteria obtained, the large-scale industrial setups are calculated.

Let us return to the example of equations (II.36) and (II.37). If we assume that the system of these equations is solved at set initial and boundary conditions, then by using this solution, we can determine the criterion F_0 by model experiments. This criterion depends on the rate of flow u, as well as the parameters D^* and β. Moreover, all the quantities are interdependent. Thus, for example, the viscosity of the medium, diameter of the grains of the sorbent and their shape, as well as the pressure drop exert an influence on the rate of flow. The quasidiffusion parameter D^* also depends on many factors: the viscosity of the medium, diameter and shape of the sorbent grains, and coefficients of molecular diffusion. The kinetic parameter β is related to many parameters determining the external and internal diffusion. It is practically impossible to establish these relationships theoretically by solution of the complex systems of equations considered above. Hence a single way remains—the experimental establishment of these relationships by model experiments. By varying the experimental conditions—the geometrical parameters (for example, grain size, column diameter, etc.), the rate of flow, concentration, etc.—and experimentally determining the criterion of similarity F_0, the empirical dependence of the criterion F_0 on various experimental parameters can be established according to the sorption dynamics. Usually all these determining parameters are also grouped in the form of dimensionless criteria of similarity, for example, the Reynolds number $Re = ud/\eta$, where d is the grain diameter, the Péclet number $Pe = ud/D_m$, where D_m is the coefficient of molar diffusion, etc. The function

$$F_0 = f(\text{Re, Pe} \ldots). \qquad (\text{II}.39)$$

may be determined experimentally. Thus, the criterion F_0 is clearly determinable, while the remaining criteria are the determining ones.

Thus, the combination of theoretical approximate solutions and the establishment of the empirical relationships among the criteria and parameters of similarity ultimately leads to practical success—provides the possibility of a practical solution of technological problems.

8. Statistical Method

A real sorbent represents a dispersed medium, permeable to a liquid or gas, in which nonequivalent sorption centers are randomly distributed—active bonds, capable of trapping atoms, molecules, or ions from the mobile phase moving through the sorbent. During the process of motion through the chromatographic column, each of the particles is successively sorbed and desorbed. The average number of events of sorption per unit length of the column depends on the summary action of the physicochemical and geometrical factors determining the kinetics, statics, and dynamics of sorption. The time of stay of a molecule in the sorbent is a random quantity; it differs for different particles. The arrangement of the sorption centers is random, while the very act of sorption is a random process for each particle. The motion of the particles is also random in character. The field of flow rates in the layer of sorbent also possesses a statistical distribution. All these statistical factors show that even when conditions of sorption equilibrium are approached, the distribution of substances at the boundaries of the chromatographic zones will possess a blurred character.

The statistical method of solving the problem of sorption dynamics, just like the phenomenological method, permits a determination of the distribution function of the substances to be sorbed in the chromatographic columns [105, 146-150, 163-165].

In the simplest case of linearity of the isotherm, it suffices to introduce two probability functions into consideration—$P_\theta(t)$ and $P_l(x)$, the physical meaning of which will consist of the following. The quantity $P_\theta(t)dt$ is the probability that the particle to be sorbed, falling from the flux into the sorbent at $t = 0$, will return to the flux in a time interval $(t, t + dt)$. Analogously, $P_l(x)dx$ gives the probability of transport of a particle over a distance $(x, x + dx)$.

The average time of stay of a particle in the sorbent θ, average transport length l, and mean square deviations for these quantities $\sigma_\theta \theta^2$ and $\sigma_l l^2$ are determined by the formulas

$$\theta = \int_0^\infty P_\theta(t)\, t\, dt, \quad \sigma_\theta \theta^2 = \int_0^\infty P_\theta(t)(t - \theta)^2\, dt; \qquad (\text{II}.40)$$

25

$$l = \int_{-\infty}^{\infty} P_l(x)\, x\, dx, \quad \sigma_l l^2 = \int_{-\infty}^{\infty} P_l(x)\,(x - l)^2\, dx.$$

(II.41)

Using the probability functions introduced and the corresponding mathematical apparatus, for example, operational calculus, we can calculate the distribution function of a substance in a chromatographic column [163-165].

A comparison of the distribution functions obtained by the statistical and phenomenological methods makes it possible to give a statistical, molecular-kinetic interpretation of the constants of the phenomenological equations of sorption dynamics.

Thus, the phenomenological and statistical methods mutually supplement one another and give a more thorough description of chromatographic processes. The relationship between the phenomenological and statistical description of the processes of sorption dynamics can be illustrated for the example of the method of lagging coordinates. From the statistical standpoint, the kinetic constant $\varkappa$ possesses the physical meaning of the average length of the "jump" of the particle to be sorbed between two successive events of sorption. The kinetic constant τ, from the statistical standpoint, can be considered as the time of the cycle, consisting of transfer of the particles to be sorbed from the flux to the sorbent and back.

The use of the statistical method in the solution of problems of sorption dynamics is not considered in this work, since it has not been the subject of the author's investigations.

CHAPTER III

THEORY OF THE FRONTAL DYNAMICS OF SORPTION
OF ONE SUBSTANCE

1. Initial Equations

Let us begin our survey of the results of the general theory of sorption dynamics and chromatography with the theory of frontal sorption dynamics and chromatography.

In this chapter, let us consider the simplest problem of sorption dynamics of one substance. We shall consider separately two systems of sorption dynamics: equilibrium (instantaneous establishment of equilibrium) and nonequilibrium. A study of sorption dynamics in an equilibrium system gives us the possibility of revealing the principles of the influence of static factors of sorption (sorption isotherm) on the sorption dynamics in the pure form. A study of the sorption dynamics in a nonequilibrium system—a more complex problem in comparison with the equilibrium sorption dynamics—permits us to elucidate the principles of the influence of kinetic factors on the sorption dynamics.

The initial system of equations of equilibrium sorption dynamics is the following:

$$\frac{\partial n}{\partial t} + \frac{\partial N}{\partial t} + u\,\frac{\partial n}{\partial x} = D^*\,\frac{\partial^2 n}{\partial x^2};\qquad (II.26)$$

$$N = f(n).\qquad (II.29)$$

As has already been noted, an inseparable component of the formulation of the problem of sorption dynamics is the setting of definite initial and boundary conditions. In rather general form, the initial and boundary conditions for the problem of frontal equilibrium sorption dynamics are the following:

$$t = 0,\ x = 0,\ n = n_0,\ N = N_0;$$
$$0 < x < x_0;\ n = \varphi(x),\ N = f(n) = f[\varphi(x)];$$
$$x > x_0,\ n = 0,\ N = 0;\qquad (III.1)$$

$$x = 0,\ t > 0,\ n = n_0,\ N = N_0;\qquad (III.2)$$

$$x = \infty,\ 0 \leqslant t \leqslant \infty,\ n = 0,\ N = 0.\qquad (III.3)$$

Let us explain the physical meaning of these conditions. The initial condition (III.1) means that at the initial moment, there is a set initial distribution of the substance to the sorbed in the column of the sorbent (initial zone), characterized by a continuous differentiable distribution function $n = \varphi(x)$ and, correspondingly, the function $N = f(n) = f[\varphi(x)]$, according to the sorption isotherm in the interval $0 \leq x \leq x_0$, where x_0 is the width of the initial zone. In order for the function $n = \varphi(x)$ not to experience any discontinuity at the inlet, let us assume that when $x = 0$, the concentration n is equal to the initial concentration of the substance in the mobile phase, i.e., $n = n_0$ and $N_0 = f(n_0)$, respectively. In the initial stage, the column of the sorbent may contain none of the substance to be sorbed at all, i.e., may be "clean." In this case, let us write the initial condition (III.1) in the following form:

$$t = 0,\ x > 0,\ n = 0,\ N = 0 \quad (\text{"clean" column}).\qquad (III.1')$$

The second condition (III.2) means, in the first place, that at the initial and any following moments of time, a constant initial concentration of the substance to be sorbed n_0 is maintained at the entrance to the column ($x = 0$) in frontal sorption dynamics, and, in the second place, that as a result of the condition of equilibrium of the sorption dynamics (instantaneous establishment of equilibrium) at the initial and any following moments of time, a constant concentration of the substance in the sorbent N_0 should be maintained at the entrance to the column, at equilibrium with the concentration n_0 according to the sorption isotherm: $N_0 = f(n_0)$.

The last equation (III.3) shows that there is an infinitely long column, and that some end of it will always be unfilled by the substance to be sorbed, i.e., will be "clean."

Let us write the initial system of equations of nonequilibrium sorption dynamics in the following form:

$$\frac{\partial n}{\partial t} + \frac{\partial N}{\partial t} + u \frac{\partial n}{\partial x} = D^* \frac{\partial^2 n}{\partial x^2}; \qquad (\text{II.26})$$

$$\frac{\partial N}{\partial t} = \psi(n, N). \qquad (\text{II.27})$$

The initial and boundary conditions for the problem of frontal nonequilibrium sorption dynamics can be set in the following way:

$$t = 0, \ \ 0 \leqslant x \leqslant x_0, \ n = \varphi(x), \ N = f(n) = f[\varphi(x)], \qquad (\text{III.1})$$

or in the case of the initial "clean" column:

$$t = 0, \ x > 0, \ n = 0, \ \ \ N = 0; \qquad (\text{III.1'})$$

$$x = 0, \ t \geqslant 0, \ n = n_0, \ N = f(t); \qquad (\text{III.4})$$

$$x = \infty, \ 0 \leqslant t \leqslant \infty, \ n = 0, \ N = 0. \qquad (\text{III.3})$$

Thus, the initial and boundary conditions for frontal nonequilibrium sorption dynamics differ from the conditions of equilibrium sorption dynamics in that the concentration of the substance in the sorbent at the entrance to the column is nonequilibrium and varies with time according to the integral equation of the sorption kinetics $N = f(t)$. At the limit, when $t \to \infty$, the concentration N will approach the equilibrium value:

$$N \to N_0 = f(n_0).$$

Let us recall that we shall always assume the presence of a sorption column "homogeneous" with respect to properties. This means that the average cross-sectional area of transport of the substance is the same for all values of the coordinate x, and all other properties of the sorbent are also standard. In the further study of the general theory of sorption dynamics and chromatography, as a rule, we shall maintain an analytical approach in the investigation, i.e., we shall simplify as much as possible the formulation of the problem, consider the simplest cases, and then, by gradually complicating the conditions, pass on to a consideration of more complex cases.

2. Equilibrium Sorption Dynamics in the Absence of Longitudinal Effects ($D^* = 0$)

Let us assume that equilibrium sorption dynamics are accomplished under ideal conditions such that the longitudinal effects of the diffusion and quasidiffusion types are negligible and can be disregarded ($D^* = 0$).

Under this simplifying condition, the system of equations describing the process of frontal equilibrium sorption dynamics takes the following form:

$$\frac{\partial n}{\partial t} + u \frac{\partial n}{\partial x} + \frac{\partial N}{\partial t} = 0; \qquad (\text{III.5})$$

$$N = f(n). \qquad (\text{II.29})$$

First let us consider the general solution of this system of equations, and then the solution of the problem at different set initial conditions (the Cauchy problem).

Let us begin the solution by eliminating the variable N in equation (III.5), using the equation of the sorption isotherm (II.29). Let us take the partial derivative with respect to time for the sorption isotherm

$$\frac{\partial N}{\partial t} = \frac{dN}{dn} \cdot \frac{\partial n}{\partial t},$$ (III.6)

where $dN/dn = f'(n)$ is the derivative of the sorption isotherm (II.29), and we shall henceforth use the abbreviated notation f'.

Substituting (III.6) into (III.5) and performing transformations, we obtain

$$(1 + f')\frac{\partial n}{\partial t} + u \frac{\partial n}{\partial x} = 0$$ (III.7)

or

$$\frac{\partial n}{\partial t} + v_n \frac{\partial n}{\partial x} = 0,$$ (III.7')

where

$$v_n = \frac{u}{1 + f'}$$ (III.8)

in some variable possessing the dimensions of velocity and depending on the concentration n. We shall explain the physical meaning of the quantity below.

Equation (III.7), according to the classification of equations in partial derivatives, is a quasilinear homogeneous equation in first order partial derivatives. Let us integrate this equation by the method of characteristics. The problem of integrating equation (III.7) is equivalent to the problem of integrating the following, so-called characteristic equation:

$$\frac{dt}{1} = \frac{dx}{v_n}.$$ (III.9)

According to the method of characteristics, the integral of equation (III.9)

$$z = x - v_n t$$ (III.10)

is the solution of equation (III.7), while the general solution of this equation represents an arbitrary, continuously differentiable function in implicit form of the integral (III.10):

$$\Phi(z) = \Phi(x - v_n t) = 0.$$ (III.11)

Let us now turn to the solution of the Cauchy problem, i.e., to a search for a solution satisfying the set initial conditions. As the initial conditions, let us take the conditions (III.1)-(III.3). The solution of the Cauchy problem is accomplished in the following way in this case. When $t = 0$, the characteristic integral of (III.10) $z = x$, and since according to the initial condition $n = \varphi(x)$, then $n = \varphi(z)$. The solution of the problem therefore will be

$$n = \varphi(x - v_n t),$$ (III.12)

where φ is a continuously differentiable function, characterizing the initial distribution of the substance in the column. Let us mention that equation (III.12) represents the functional dependence of n on x and t in implicit form, since, according to (III.8), v_n depends on the variable n.

Equation (III.12) does not permit the expression of the dependence of the concentration n on x and t in explicit form, when it is set by such a general form. However, it can easily be transformed to the explicit form with respect to the variable x (or t). For this, let us first solve equation (III.12) with respect to the variable

$$x - v_nt = \bar{\varphi}(n),$$

(III.13)

where $\bar{\varphi}$ is a reciprocal function with respect to the function φ.

From (III.13) we obtain

$$x = \bar{\varphi}(n) + v_nt.$$

(III.14)

When $t = 0$, $x = \bar{\varphi}(n)$ gives us the initial distribution of the substance, expressed in terms of a reciprocal function of the function $n = \varphi(x)$.

Equation (III.14) is more convenient for further analysis of the form of the solution of the problem posed. Let us emphasize especially that this solution is correct only in the case when there is a continuous spectrum of concentrations in the front of sorption dynamics. From the physical standpoint, equation (III.14) is the equation of motion of the concentration points of the sorption dynamics front. It permits a consideration of the distribution of the substance to be sorbed along the column of the sorbent for various moments of time. The physical meaning of the quantity v_n consists of the fact that it is the value of the rate of displacement along the column of sorbent of a set concentration n.

The rate of motion of the concentration points of the front of sorption dynamics v_n, according to (III.8), depends on the derivative of the sorption isotherm. Each concentration point corresponds to a definite value of the derivative of the sorption isotherm. Consequently, each concentration point of the continuous front of sorption dynamics, according to (III.8), should move at a characteristic constant rate, unambiguously determined by the derivative of the sorption isotherm. The solution of the problem under consideration and formula (III.8) were first obtained by E. Wicke [233-235]. The formula of motion of the concentration points of the front of equilibrium sorption dynamics (III.8) in the pure form reveals the action of one of the factors of sorption dynamics— the factor of the sorption isotherm—and expresses one of the most important laws of sorption dynamics. Henceforth we shall call formula (III.8) Wicke's law. The solution of the equation of sorption dynamics (III.7) considered is not unique. This equation is also satisfied by the obvious solution: $n = $ const, and, correspondingly, $N = f(n) = $ const. The physical meaning of this solution consists of the fact that if there is one—a unique concentration of the substance in the column of the sorbent, then it should be maintained during motion along the column. The question of the rate of motion of such a unique concentration point will be considered later.

Let us mention also that the method of solving the equation of sorption dynamics that we used (method of characteristics) is likewise not a unique method of solution.

Other methods can also be used for the solution of differential equations in partial derivatives, for example, the method of replacement of variables. Let us consider it briefly. The problem consists of selecting a new variable such that the equation in partial derivatives might be transformed to the usual differential equation with its aid. The new variable for the problem under consideration can be selected on the basis of the following physical prerequisites.

According to the initial condition (III.1), at the initial moment $t = 0$, an initial distribution of the substance expressed by the function $n = \varphi(x)$ or by the reciprocal function $x = \bar{\varphi}(n)$ exists in the column. The latter function gives the value of the x coordinate for the given concentration point n. This concentration point should move along the column during the process of sorption dynamics with some velocity v_n, as yet unknown to us. Then the equation of the motion of this point at $t > 0$ will be

$$x = v_nt + \bar{\varphi}(n).$$

This equation defines the coordinate of displacement of a set concentration point at a set initial distribution at any moment of time t. By solving the last equation with respect to n, we already obtain the solution that we know, (III.12): $n = \varphi(x - v_nt)$. It permits us to take a new variable $z = x - v_nt$ to replace the variables x and t and to express the concentration n as a function of this new variable: $n = \varphi(z)$. The transformation of the partial derivatives in (III.7) through the first variable gives

$$\frac{\partial n}{\partial t} = -v_n \frac{d\varphi}{dz}\ ; \quad \frac{\partial n}{\partial x} = \frac{d\varphi}{dz}.$$

Thus, with the aid of a new varibale, equation (III.7) is converted to a normal differential equation:

$$-v_n (1 + f') \frac{d\varphi}{dz} + u \frac{d\varphi}{dz} = 0,$$

from which

$$u - v_n (1 + f') = 0 \text{ and } \frac{d\varphi}{dz} = 0.$$

This leads us again to two solutions of equation (III.7). The first solution is $n = \varphi(x-v_n t)$, where $v_n = u/(1 + f')$, and the second solution, which follows from the condition $d\varphi/dz = 0$, is $n = \varphi(z) = \text{const}$. Thus, the method of replacing variables leads to the already known solutions.

Let us turn to a consideration of the physical results that follow from the solutions obtained. As has already been mentioned, Wicke's law (III.8) reveals the influence of the type of isotherm on the character of the sorption dynamics. In a study of sorption phenomena, three types of sorption isotherms are most often encountered: convex, concave, and linear (Fig. 2). In accord with this, three systems of sorption dynamics, pertaining to convex, concave, and linear isotherms, can be distinguished. Let us consider each of these systems of sorption dynamics individually.

The Convex Isotherm $f'(n_1) > f'(n_2)$ when $n_1 < n_2$. In the case of a convex sorption isotherm, the concentration points with greater concentrations, according to Wicke's law (III.8), will move at greater speeds, while the points with lower concentrations will move at lower speeds: when $n_1 < n_2$ we have $v_{n1} < v_{n2}$. Hence, if there were a blurred front of the substance in the column of the sorbent at the initial moment [see conditions (III.1)-(III.3)], then compression of the front should occur during the process of frontal equilibrium sorption dynamics according to Wicke's law.

Let us consider the physical essence of this phenomenon in greater detail. Actually, we need to determine the region of the variable arguments x and t in which the solution (III.14) will exist. In other words, we must determine the region of action of Wicke's laws under the conditions of a system of sorption dynamics with a convex isotherm. It is evident from solution (III.14) that the motion and distribution of the substance in the process of sorption dynamics depends not only on the sorption isotherm, but also on the initial distribution. The functions of the initial distribution $\varphi(x)$ or $\varphi'(n)$ can be most varied—any set continuous distribution of the substance.

To illustrate certain characteristic principles of sorption dynamics in the case of a convex isotherm, let us set a number of these initial distributions. Let us assume, for example, that the function of the initial distribution $\bar{\varphi}(n)$ is a continuous, monotonically decreasing function, taking values $n = n_0$ when $x = 0$ and $n = 0$ when $x = x_0$. According to Wicke's law (III.8), the point $n = n_0$ will move at the greatest rate v_{n0} in the case of a convex isotherm, while the point n_0 will move at the least rate v_0. The points of intermediate concentrations $0 < n < n_0$ will move at intermediate velocities: $v_0 < v_n < v_{n0}$.

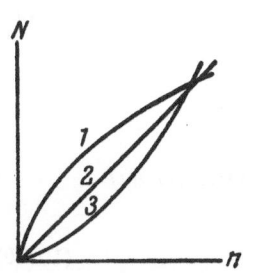

Fig. 2. Three types of sorption isotherms. 1) Convex; 2) linear; 3) concave.

Let us undertake the following task—to determine the form that the function of the initial distribution $\bar{\varphi}(n)$ should take for all the concentration points of the deformed, compressed front to reach the same coordinate x at the same time. This will be the unique critical moment of sorption dynamics, at which the blurred front disappears, and the distribution function of the substance along the column will be converted from a continuous differentiable function to a discontinuous nondifferentiable function—the front with blurring is converted to a front with a straight break. At this moment, all the concentrations disappear, with the exception of the unique concentration n_0. Let us denote the co-

31

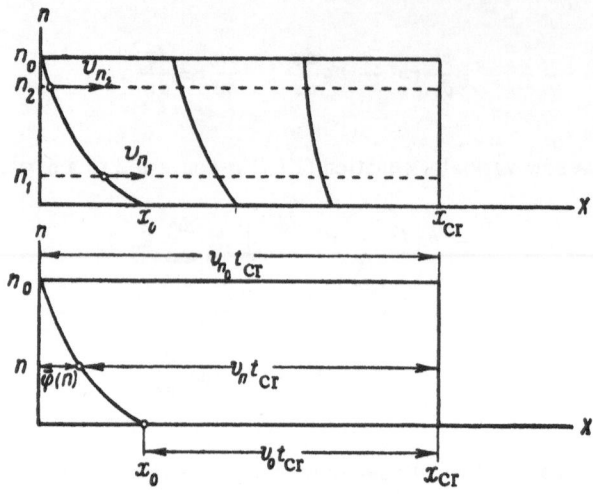

Fig. 3. Equilibrium sorption dynamics of one substance
in the case of a convex isotherm. For calculation of the
distribution in the initial zone, in which a simultaneous
complete "rectification" band of the blurred front should
occur. Velocity $v_{n1} < v_{n2}$; velocity $v_0 < v_{n0}$.

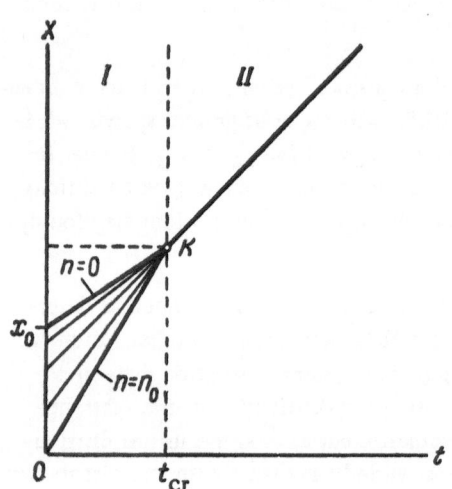

Fig. 4. Graphs of the motion of the concentration points of a front of equilibrium sorption dynamics in the case of a convex isotherm in coordinates (x, t). I) Stage of compression of the front; II) stage of motion of
the stationary front; K) critical point of
transition from the first to the second stage
of sorption dynamics.

ordinate at which a front with break appears as x_{cr} and
the moment of appearance of such a front itself as t_{cr}.

The pehnomenon of compression of the front and its
conversion to a front with break described is schematically
shown in Fig. 3.

According to (III.13), the function of the initial distribution $\bar{\varphi}(n)$ sought should be determined according to the
equation (see Fig. 3)

$$\bar{\varphi}(n) = x_{cr} - v_n t_{cr}. \qquad (III.15)$$

To determine it, we must find the critical coordinates x_{cr}
and t_{cr}. They can easily be found from the following system
of equations:

$$x_{cr} = v_{n_0} t_{cr}, \qquad (III.16)$$

$$x_{cr} - x_0 = v_0 t_{cr}, \qquad (III.17)$$

from which

$$x_{cr} = \frac{x_0 v_{n_0}}{v_{n_0} - v_0}, \qquad (III.18)$$

$$t_{cr} = \frac{x_0}{v_{n_0} - v_0}. \qquad (III.19)$$

Substituting (III.18) and (III.19) into (III.15), we obtain the function of the initial distribution sought:

$$x_n = \bar{\varphi}(n) = \frac{x_0 v_{n_0}}{v_{n_0} - v_0} - v_n \frac{x_0}{v_{n_0} - v_0}. \qquad (III.20)$$

In the case of the initial distribution of the substance to be sorbed described by function (III.20), the distribution of the substance at the stage of compression of the front will be described by the equation

$$x = \frac{x_0 v_{n_0}}{v_{n_0} - v_0} - v_n \frac{x_0}{v_{n_0} - v_0} + v_n t \qquad \text{(III.21)}$$

where $x \leq x_{cr}$ and $t \leq t_{cr}$.

The process of sorption dynamics in the case of a convex isotherm can be characterized graphically at the stage of compression of the front in the coordinate system (x, t). Figure 4 presents curves of the motion of various concentration points of the initial blurred front in the coordinates (x, t). In the geometrical interpretation, the dependence of the concentration n on x and t is depicted by a surface in three-dimensional space (n, x, t). Consequently, the curves of the motion of the points of the front of sorption dynamics, reflecting the dependence of the displacement on the time t, is a projection of the points of this surface onto the plane of the coordinates (x, t). The segment Ox_0 on Fig. 4 is a projection of the initial front onto the OX-axis. A definite concentration according to equation (III.20) corresponds to each point x on this segment.

According to Wicke's law, the graphs of the motion of the concentration points should be represented by straight lines, with slopes differing, depending on $f'(n)$. In the case under consideration, the concentration n_0 is found at the origin at the initial moment. The graph of the motion of this point will be represented by a straight line passing through the origin and possessing the greatest slope. The graphs of the motion of the other intermediate concentration points will begin at the segment Ox_0. These are straight lines, each of which possesses a slope smaller than the preceding. The last point—the concentration point $n = 0$ when $x = x_0$—will move at the lowest rate. A characteristic feature of the case under consideration is the fact that we selected an initial distribution of the substance such that all the concentration points of the front will converge at one coordinate x_{cr} at a definite critical moment t_{cr}. This will be the moment of disappearance of the blurred front and the end of the action of Wicke's law. Thus, in Fig. 4 the region of the action of Wicke's law is bounded by the region of the triangle Ox_0K, i.e., the region of deformation of the blurred front.

From the moment of disappearance of the blurred front and the appearance of the front with break of the unique concentration n_0, a new stage in sorption dynamics begins, which will be characterized by new boundary conditions. Let us write these new boundary conditions:

$$t = t_{cr}, \ x > x_{cr}, \ n = 0, \ N = 0; \qquad \text{(III.22)}$$

$$x = x_{cr}, \ t \geqslant t_{cr}, \ n = n_0, \ N = N_0; \qquad \text{(III.23)}$$

$$x = \infty, \ t_{cr} \leqslant t \leqslant \infty, \ n = 0, \ N = 0. \qquad \text{(III.24)}$$

The conditions of the problem under consideration contain the assumption that there are no kinetic and longitudinal effects, and there is no action of any perturbing factors of blurring of the front in the front of sorption dynamics. Then, according to the second solution of equation (III.7), n = const, the concentration $n = n_0$ that appears in the breakoff front should remain the unique concentration during the entire further process of sorption dynamics. In this case, of course, the equilibrium concentration in the sorbent will also remain unchanged: $N = N_0 = f(n_0)$. To determine the rate of motion of the breakoff front with concentration n_0, we must know not the differential, but the integral form of the equation of the material balance, since in this case we are dealing not with a continuously differentiable function, but with a constant.

If we construct the cross section of the column at the level $x = x_{cr}$, then during a virtual interval of time, an amount of the substance equal to $un_0 \delta t$ should flow through this cross section. But this amount of the substance, passing through this cross section, is distributed between the sorbent and mobile phase as a result of sorption dynamics. In this case, no changes should occur in the concentrations in the front—the previous concentrations $n = n_0$ and $N = N_0$ are preserved, while the front itself will possess a break. The only result of the process will be the motion of the front of sorption dynamics at some as yet unknown velocity over some distance. To determine this velocity, let us compile the explicit equation of the material balance

$$un_0 \, \delta t = v \, (n_0 + N_0) \, \delta t, \qquad \text{(III.25)}$$

where v is the rate of motion of the front.

33

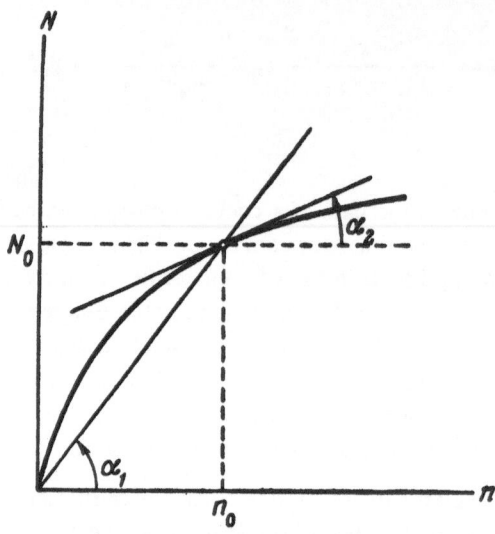

Fig. 5. Angles of inclination of the secant and
tangent for a convex isotherm.

On the right in this equation is the sum of the amounts of the substances in the sorbent and mobile phase in the cross section of the column at the level x_{cr}.

The velocity sought can be obtained from equation (III.25):

$$v = u \, \frac{n_0}{n_0 + N_0}. \tag{III.26}$$

This equation shows that, in the first place, the rate of motion of the front with the unique concentration $n = n_0$ is a constant (of course when $u = $ const). Thus, with the critical moment of disappearance of blurring of the front begins the steady-state stage of motion of the breakoff front at a constant rate, determined by the formula (III.26). This velocity will differ from the velocity v_{n_0} according to Wicke's law (III.8). If we turn to the graph of the convex sorption isotherm (Fig. 5), then we see that at the stage of compression of the blurred front, the rate of motion of the concentration n_0 and the breakoff front at the steady-state stage is determined by the ratio N_0/n_0 (slope of the secant). The velocity of the steady-state breakoff front v will be less than the velocity v_{n_0}, which possessed a concentration point at the stage of compression of the blurred front: $v < v_{n_0}$.

Correspondingly, on the graph of motion of the concentration points (see Fig. 4), the slope of the straight line characterizing the motion of the steady-state breakoff front at the stage $t > t_{cr}$ will be smaller than the slope of the straight line characterizing the motion of the concentration $n = n_0$ at the preceding stage.

Thus, at the stage of motion of the steady-state front with break, the distribution of the substance is the following:

$$\left. \begin{array}{l} t > t_{cr}, \ x_{cr} \leqslant x \leqslant x_{cr} + v \, (t - t_{cr}), \ n = n_0, \ N = N_0; \\ x > x_{cr} + v \, (t - t_{cr}), \ n = 0, \ N = 0, \end{array} \right\} \tag{III.27}$$

where v is the rate of motion of the steady-state front according to formula (III.26).

We considered one of the special cases of deformation of a blurred front in a system of sorption dynamics with a convex isotherm. However, even this particular example shows that the processes of dynamics can possess a stepwise character.

In other cases, when the function of the initial distribution differs from the distribution (III.20), the process of rectification of the front will be more complex. However, no matter what the initial distribution $\bar{\varphi}(n)$, on the strength of the action of Wicke's law, at some stage of the sorption dynamics there should be complete

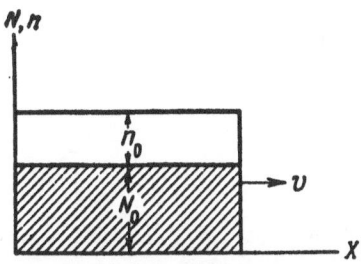

Fig. 6. Equilibrium sorption dynamics of one substance at the initial and boundary conditions (III.28)-(III.30). Shaded area—composition of the sorbent; unshaded area—composition of the mobile phase.

rectification of the front, and the unique concentration n_0 in equilibrium with the concentration N_0 is correspondingly established. From this moment begins the steady-state stage of sorption dynamics—the front with breakoff begins to move at a constant speed, according to formula (III.26). Conditions analogous to the conditions (III.22)-(III.24) can be set as the original, initial, and boundary conditions.

Let us assume that at the initial moment the column was "clean," i.e., that it did not contain the substance to be sorbed, while the substance with a concentration n_0 began to be delivered at the entrance to the column. In view of the assumed equilibrium character of the sorption dynamics at the entrance to the column, at the initial moment an instantaneous equilibrium should be established, while the concentration in the sorbent will be equal to $N_0 = f(n_0)$. Let us write the initial and boundary conditions for this case in mathematical form:

$$t = 0, \ x > 0, \ n = 0, \ N = 0; \tag{III.28}$$

$$x = 0, \ t > 0, \ n = n_0, \ N = N_0 \ \text{(instantaneous equilibrium)}; \tag{III.29}$$

$$x = \infty, \ 0 \leqslant t \leqslant \infty, \ n = 0, \ N = 0. \tag{III.30}$$

These conditions show that at the initial moment a distribution of the substance characterized by the presence of the unique concentration n_0 in equilibrium with the concentration N_0 is observed at the entrance to the column (x = 0).

As was shown above, the unique concentration n_0 in equilibrium with the concentration should move at the same speed. The original break in the distribution function will be preserved—the steady-state front of sorption dynamics should move along the column at a constant rate, determined by formula (III.26) (see Fig. 6).

Let us introduce a quantity that we shall call the partition coefficient:

$$h = \frac{n_0}{N_0} = \frac{n_0}{f(n_0)} \ . \tag{III.31}$$

Then the formula for the rate of motion of the steady-state front (III.26) can be represented in the following form:

$$v = u \ \frac{h}{1+h}, \tag{III.32}$$

i.e., the rate of motion of the steady-state front depends only on the value of the partition coefficient at a set constant rate of flow.

Hence, for the initial and boundary conditions (III.28)-(III.30), we have the following solution:

$$0 \leqslant x \leqslant vt, \ n = n_0, \ N = N_0; \tag{III.33}$$

$$x > vt, \ n = 0, \ N = 0. \tag{III.34}$$

If some sort of random perturbation occurs during the process of equilibrium sorption dynamics, leading to some blurring of the front, then, according to Wicke's law, the factor of influence of the convexity of the sorption isotherm—factor of compression of the front—goes into action, and the blurring soon disappears.

Hence, the convexity of the isotherm is a factor of compression and stabilization of the front of sorption dynamics.

The formula for the rate of motion of the unique concentration point of equilibrium sorption dynamics (III.26) or (III.32), was first obtained by J. Wilson [241]; hence, we shall henceforth call it the formula of the

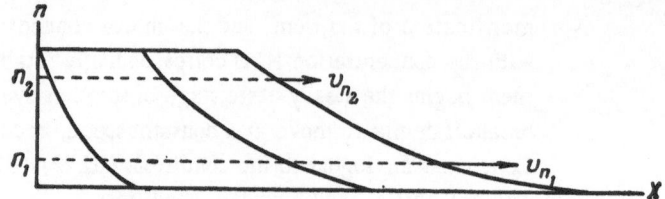

Fig. 7. Equilibrium sorption dynamics in the case of a concave isotherm. Progressive blurring of the front: $v_{n1} > v_{n2}$.

Wilson law. The laws of Wicke and Wilson exhaustively describe the process of equilibrium sorption dynamics when $D^* = 0$. The shape of the distribution of the substance in the column at any stage of equilibrium sorption dynamics can be predicted with their aid when there are no longitudinal effects of blurring.

Let us turn to a consideration of the characteristics of sorption dynamics in the case of a concave isotherm.

The Concave Isotherm $f'(n_1) < f'(n_2)$ when $n_1 < n_2 < n_0$. In the case of a concave sorption isotherm, the process of deformation of the blurred front according to Wicke's law is opposite to that which we were considering just before. In this case, according to (III.8), the points of the blurred front with lower concentration will move at greater speed, while the points with higher concentration will move more slowly, i.e., there will be a progressive blurring of the front (Fig. 7). Each point of the front will move along the column with its own characteristic rate constant according to formula (III.8). The distribution of the substance in the process of sorption dynamics, as it follows from the solution (III.14), also depends on the function of the initial distribution of the substance $\bar{\varphi}(n)$.

Let us consider one particular case of the initial distribution of the substance. Let us assume that the initial zone of the substance is very narrow, i.e., $x_0 \approx 0$, or, in other words, the width of the primary zone is considerably smaller than the distances through which the concentration points of the blurring front subsequently pass: $\bar{\varphi}(n) << v_n t$. Let us also assume that the distribution function is continuous and differentiable on the area of the initial zone.

According to the conditions adopted, at any $0 < n < n_0$, the quantity $\bar{\varphi}(n)$ can be neglected in the solution of (III.14) in comparison with $v_n t$. In the process of sorption dynamics, in the case of a concave isotherm, each concentration point will move at its own characteristic speed, beginning practically from the entrance to the column $x \approx 0$. Hence, the dynamic curve of the distribution of the substance in this case will obey the following equation, which follows from the solution (III.14):

$$x = v_n t \tag{III.35}$$

or

$$x = \frac{ut}{1 + f'(n)} . \tag{III.36}$$

In general, when the initial distribution of the substance cannot be disregarded, the complete form of the solution (III.14) should be used to describe the dynamics of equilibrium sorption in the case of a concave isotherm.

If we arbitrarily set some concentrations n_1 and n_2, so that $0 < n_1 < n_2 < n_0$, and define the width of the front as the distance $\delta_x = x_{n1} - x_{n2}$, then it follows from (III.14) that:

$$\delta_x = x_{n_1} - x_{n_2} = [\bar{\varphi}(n_1) - \bar{\varphi}(n_2)] + (v_{n_1} - v_{n_2}) t, \tag{III.37}$$

i.e., the width of the front increases linearly with the time of the process.

Hence, the concavity of the sorption isotherm is the factor of blurring of the front of sorption dynamics. This result follows from Wicke's law.

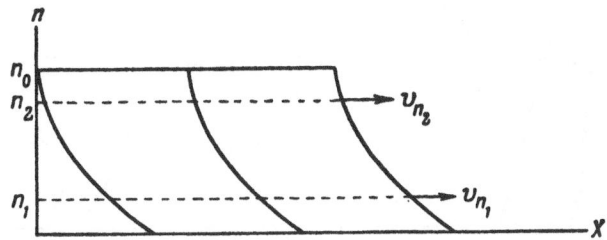

Fig. 8. Equilibrium sorption dynamics in the case of a linear isotherm. Conservation of blurring of the front: $v_{n_1} = v_{n_2}$.

The Linear Isotherm $f'(n_1) = f'(n_2) = \text{const.}$ In the case of a linear isotherm, the derivative of the isotherm is a constant. Hence, according to Wicke's law, all the points of the blurred front should move at a constant speed. For the linear isotherm, the derivative of the isotherm should be numerically equal to the ratio N_0/n_0:

$$f'(n) = \frac{N_0}{n_0} = \frac{1}{h}. \qquad (III.38)$$

Hence the velocity of the concentration points of the blurred front in the case of a linear isotherm will be

$$v = \frac{u}{1 + f'(n)} = u \frac{h}{1 + h}. \qquad (III.39)$$

The formula obtained is analogous to the formula of the velocity of the steady-state front with a break (III.26) or (III.22). Hence, in the process of equilibrium sorption dynamics when $D^* = 0$ and the isotherm is linear, a system of parallel transport of the initial front is set up (Fig. 8). The linearity of the sorption isotherm is the factor of conservation of blurring of the front of sorption dynamics.

Although in the case of a convex isotherm, any random perturbation arising in the front can be eliminated, while in the case of a concave isotherm, on the contrary, it becomes the cause of a progressive blurring, in the case of a linear isotherm, blurring that arises randomly in the front will be preserved in the case of equilibrium sorption dynamics.

3. Equilibrium Sorption Dynamics under the Action of Longitudinal Effects ($D^* \neq 0$)

Let us consider the theory of equilibrium sorption dynamics of one substance under the action of longitudinal (diffusion and quasidiffusion) factors of blurring of the front ($D^* \neq 0$). In this case the sorption dynamics will be described by the system of equations (II.26) and (II.28). Eliminating the variable N from equation (II.26) through the sorption isotherm (II.28), we obtain

$$\frac{\partial n}{\partial t} + v_n \frac{\partial n}{\partial x} = D^* \frac{\partial^2 n}{\partial x^2}. \qquad (III.40)$$

According to the classification of equations in partial derivatives, this equation is an equation in second-order partial derivatives of the parabolic type. No general solution of this equation has been found. However, an approximate picture of sorption dynamics under the conditions considered can be obtained on the basis of a physical analysis of these conditions. The first factor that should be mentioned is the action of the factor of blurring of the front in the process of sorption dynamics ($D^* \neq 0$). As soon as the substance begins to enter the column ($x > 0$), blurring of the front begins immediately. Thus, under no conditions is the penetration of the front of sorption dynamics with a break in the distribution function possible in the process of sorption dynamics under the action of longitudinal effects. But the existence of a blurred front is necessary and sufficient for the action of Wicke's law, reflecting the influence of the sorption isotherm on the course of sorption dynamics. Hence, the second operating factor of sorption dynamics can be considered to be the influence of the type of sorption isotherm.

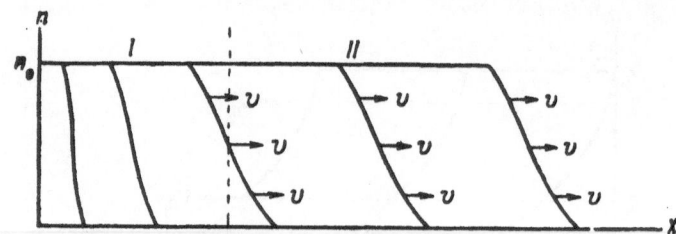

Fig. 9. Equilibrium sorption dynamics in the case of a convex isotherm under the action of longitudinal effects. I) Stage of front formation; II) stage of its parallel transport (steady-state front).

Hence, just as in the preceding section, we should consider the problem of equilibrium sorption dynamics when $D^* \neq 0$ for three systems: for convex, concave, and linear isotherms. Let us use conditions of the type of (III.28–III.30) as the initial and boundary conditions. These conditions mean that first there is a clean column of the sorbent, while the substance being sorbed, with concentration n_0, is delivered at the entrance to the column. In view of the conditions of equilibrium of the sorption dynamics at the initial moment, instantaneous equilibrium should be established, and the concentration of the substance in the sorbent at $x = 0$ will be $N_0 = f(n_0)$. Since the second derivative $\partial^2 n / \partial x^2$ enters into equation (III.40), we must know the values of the first derivative at the boundaries of the coordinate x. Since the values of the quantity n are constant at the boundaries of the coordinate x (III.29), (III.30), then at these boundaries we also have $\partial n / \partial x = 0$.

Convex Isotherm. The system of sorption dynamics in the case of a convex isotherm is characterized by the fact that two factors operate in opposite directions—the factor of blurring of the front (longitudinal diffusion and quasidiffusion effects) and the factor of compression of the front (convexity of the sorption isotherm and action of Wicke's law). On the basis of this representation of the action of two oppositely directed factors, we can advance the physical prerequisite that a mutual "equilibration" of these factors should occur at some asymptotic stage of the sorption dynamics. The blurring of the front will be compensated for by its compression, resulting in stabilization of the front, and the establishment of motion of the steady-state front, which moves along the column at a constant speed in a system of parallel transport. This representation of the conditions of emergence of a steady-state front under the action of blurring factors was first advanced by Ya. B. Zel'dovich [63] and further developed by O. M. Todes [139]. The physical prerequisite of the possibility of formation of a steady-state front is the base for formulation of the problem of seeking the asymptotic solution, i.e., determination of the distribution functions of the substance for the stage of the steady-state front. Let us note that here we are again enlisting the representation of a stepwise character of the process of sorption dynamics. In the case under consideration, two stages of sorption dynamics can be arbitrarily distinguished: the stage of formation of the steady-state front and the stage of its parallel transport (Fig. 9).

To evaluate the time of formation of the steady-state front, or the time of transition from one stage to another, we must have the complete solution of the problem of sorption dynamics. However, it does not exist. Hence, at the present time we can formulate only an approximate physical condition for evaluating the beginning of the stage of the steady-state front: the steady-state stage sets in in practice when the blurred front has traveled a sufficiently long distance along the column for the width of the front to be considerably smaller than the width of the region of saturation of the sorbent (see Fig. 9). From the theoretical standpoint, on the other hand, the stage of the steady-state front is an asymptotic stage, i.e., theoretically it sets in when $t \rightarrow \infty$. Let us use the method of propagated waves—the method of Dalambert [15, 80, 138]—to determine the distribution function of the system at the stage of the steady-state front. Actually, a steady-state front of sorption dynamics, propagated at a constant rate, can be considered as some wave. The general solution of the differential equation (II.26) for the stage of motion of a steady-state front, according to the method of propagated waves, takes the following form:

$$n = \varphi(x - vt), \qquad (III.41)$$

where φ is some function that can be determined for set concrete initial, boundary, and other conditions of the problem; v is the rate of motion of the steady-state front, which should also be determined on the basis of the conditions of the problem.

The physical meaning of the solution (III.41) consists of the fact that the equation of any steady-state wave can be written in a system of coordinates moving together with the wave. In this case, we can pass from a system of two coordinates (x, t) to a one-coordinate system $z = x - vt$, in which the contour of the wave will be described as some displacement z from the origin $z = 0$, moving at a velocity v. The position of this new "traveling" origin in the previous system of coordinates will be determined by the obvious condition $z = 0$ for the point $\bar{x} = vt$. The contour of the front is characterized by the displacements $z(n)$, situated to the right and to the left of $z = 0$, i.e., in the direction of positive and negative values of z within limits $z \to +\infty$ and $z \to -\infty$. Hence, the solution (III.41) permits us to introduce a new variable

$$z = x - vt, \tag{III.42}$$

with the aid of which we can establish the contour of the steady-state wave of sorption dynamics. Substituting the variable $z = x - vt$ into (III.41), we obtain

$$n = \varphi(z), \tag{III.43}$$

from which it follows that the new variable z is a function of the concentration n:

$$z = \bar{\varphi}(n), \tag{III.44}$$

where $\bar{\varphi}$ is reciprocal with respect to the function φ.

Thus, the general solution of the problem of determining the steady-state front can be expressed by the following equation:

$$x = vt + z(n). \tag{III.45}$$

Let us mention once again that the solution (III.41) or, which is the same, (III.45) is a general equation of any steady-state wave, propagated at a constant speed, independent of the concrete physical factors leading to its formation.

The concrete expressions of the value of the velocity v and the function $z(n)$ are determined in accord with the set physical conditions of the process. Hence, we must seek the value of the velocity v and the function $z(n)$ for the problem of frontal equilibrium sorption dynamics pased when $D^* \neq 0$ in the case of a convex isotherm. The initial system of equations is represented by equations (II.26), (II.29). Let us also write the original, initial, and boundary conditions of the problem:

$$t = 0, \quad x > 0, \quad n = 0, \quad N = 0; \tag{III.46}$$

$$x = 0, \quad t \geqslant 0, \quad n = n_0, \quad N = N_0, \quad \frac{\partial n}{\partial x} = 0; \tag{III.47}$$

$$x = \infty, \quad 0 \leqslant t \leqslant \infty, \quad n = 0, \quad N = 0, \quad \frac{\partial n}{\partial x} = 0. \tag{III.48}$$

Since we are considering the asymptotic stage of sorption dynamics $(t \to \infty)$, then the initial condition (III.46) drops out. For the new variable $z = x - vt$, the boundary conditions (III.47) and (III.48) take the following form:

$$z = -\infty, \quad n = n_0, \quad N = N_0, \quad \frac{dn}{dz} = 0; \tag{III.49}$$

$$z = +\infty, \quad n = 0, \quad N = 0, \quad \frac{dn}{dz} = 0. \tag{III.50}$$

The physical meaning of these conditions consists of the fact that to the right of $z = 0$, when $z \to +\infty$, a region of the sorbent not yet reached by the substance to be sorbed always exists, while to the left of

$z = 0$, when $z \to -\infty$, lies a region where saturation of the sorbent by the substance to be sorbed is noted. Since the concentration n is a function of the unique variable z, the partial derivative $\partial n / \partial x$ is replaced by the usual derivative dn/dz.

Let us replace the variables in the equation of the balance of sorption dynamics (II.26), using the substitution (III.42). Let us first transform the partial derivatives:

$$\left.\begin{array}{ll} \dfrac{\partial n}{\partial t} = \dfrac{dn}{dz} \cdot \dfrac{\partial z}{\partial t} = -v\dfrac{dn}{dz}, & \dfrac{\partial N}{\partial t} = -v\dfrac{dN}{dz}; \\[2mm] \dfrac{\partial n}{\partial x} = \dfrac{dn}{dz} \cdot \dfrac{\partial z}{\partial x} = \dfrac{dn}{dz}, & \dfrac{\partial^2 n}{\partial x^2} = \dfrac{d^2 n}{dz^2}. \end{array}\right\} \qquad (\text{III.51})$$

Substituting the new expressions for the derivatives into the equation of the balance (II.26), we obtain

$$-v\frac{dn}{dz} + u\frac{dn}{dz} - v\frac{dN}{dz} = D^{\bullet}\frac{d^2 n}{dz^2}. \qquad (\text{III.52})$$

After integration of equation (III.52) with respect to z, we shall have

$$D^{\bullet}\frac{dn}{dz} = (u - v)\,n - vN + C, \qquad (\text{III.53})$$

where C is the integration constant. This constant can be determined on the basis of the boundary condition (III.50). Since the solution sought should satisfy this boundary condition, then, substituting the values of the variables according to the condition (III.50), we obtain the value of the constant $C = 0$. Hence, in place of (III.53), we can write

$$D^{\bullet}\frac{dn}{dz} = (u - v)\,n - vN. \qquad (\text{III.54})$$

On the other hand, the solution (III.54) obtained should also satisfy the boundary condition (III.49). Using this boundary condition, we obtain from (III.54)

$$(u - v)\,n_0 - vN_0 = 0, \qquad (\text{III.55})$$

from which the velocity of the steady-state front sought can be found:

$$v = u\frac{n_0}{n_0 + N_0} = u\frac{h}{1 + h}. \qquad (\text{III.56})$$

Thus, the rate of motion of the blurred steady-state front when $D^{\bullet} \neq 0$ is determined according to the same formula as the rate of motion of the front with a direct break when $D^{\bullet} = 0$ [see formulas (III.26) and (III.32)].

To determine the function z(n), which enters into the general equation for the steady-state front (III.45), in other words, to determine the equation of the contour of the steady-state front, let us substitute the equation of the sorption isotherm (II.29) into equation (III.54):

$$D^{\bullet}\frac{dn}{dz} = (u - v)\,n - vf(n). \qquad (\text{III.57})$$

Solving this differential equation with respect to z and using the formula for the velocity of the steady-state front (III.56), we obtain

$$z(n) = \frac{D^{\bullet}(1 + h)}{u}\int\frac{dn}{n - hf(n)} + C, \qquad (\text{III.58})$$

where C is a new integration constant. The formula (III.58) obtained is the equation of the contour of the steady-state front for the concentration n.

The equation of the contour of the steady-state front can also be derived for the concentration N.

This equation will possess just as general a form as (III.45):

$$x = vt + z\,(N),\qquad\text{(III.59)}$$

where the concentration N is related to the concentration n by the equation of the sorption isotherm $N = f(n)$. An analogous expression can also be written for the front of the summary concentration n + N:

$$x = vt + z\,(n + N).\qquad\text{(III.60)}$$

The physical meaning of the identity of equations (III.45), (III.59), and (III.60) consists of the fact that under the conditions of a steady-state front, the interrelated equilibrium concentrations n and N, as well as their sum n + N, move at a constant velocity v, and at a given moment of time they should possess the same coordinate of displacement x. From this the equation of the displacement z in the moving system of coordinates also follows:

$$z\,(n) = z\,(N) = z\,(n + N).\qquad\text{(III.61)}$$

The function $z(N)$ can be obtained from equation (III.58) by substituting into it the equation of the sorption isotherm in the form of the reciprocal function $n = F(N)$:

$$z\,(N) = \frac{D^{*}\,(1 + h)}{u}\int\frac{F'\,(N)\,dN}{F\,(N) - hN} + C.\qquad\text{(III.62)}$$

Let us denote the first members in the right-hand portions of equations (III.58) and (III.62) as $z_0(n)$ and $z_0(N)$, respectively:

$$z\,(n) = z_0\,(n) + C;\qquad\text{(III.63)}$$

$$z\,(N) = z_0\,(N) + C.\qquad\text{(III.64)}$$

In equations (III.63) and (III.64), representing an abbreviated notation of equations (III.58) and (III.62), the functions $z_0(n)$ and $z_0(N)$ are identical. But since $z(n) = z(N)$, then the integration constants in equations (III.58) and (III.62) are equal.

For a definitive solution of the problem, i.e., to obtain the equation of motion of the stationary front for concentrations n and N, we must determine the integration constant in expressions (III.58) and (III.62). For this purpose, let us return once again to a consideration of the general form of the equation of the stationary front (III.45).

It was shown above that when $D^{*} \neq 0$, the blurred stationary front moves at the same velocity v as the stationary front with a straight break when $D^{*} = 0$ [see (III.56) and (III.32)]. If $D^{*} = 0$, then according to the Wilson law, the front of the substance is propagated in the column at a velocity v with a straight break, and the coordinate of this break $\bar{x} = vt$. Thus, the coordinate $\bar{x} = vt$ of a front with a break when $D^{*} = 0$ coincides with the beginning of the reading of z = 0 in the new coordinate system, moving at a velocity v together with the blurred stationary front at $D^{*} \neq 0$. This comparison of the two solutions shows that the coordinate $\bar{x} = vt$, or z = 0, which is the same, possesses the character of a singular point, on both sides of which diffuse blurring of the front occurred (Fig. 10).

In this case, blurring of the front should be accomplished in such a way that the law of conservation of matter is observed. This means that the areas under the distribution curves of the substance in the column (see Fig. 10) for the case $D^{*} = 0$ (front with break) and for the case $D^{*} \neq 0$ (blurred front) should be equivalent. The indicated condition, which follows from the law of conservation of matter, can be written in the following mathematical form:

$$\int_{0}^{n_0} x\,dn + \int_{0}^{N_0} x\,dN = (n_0 + N_0)\,vt.\qquad\text{(III.65)}$$

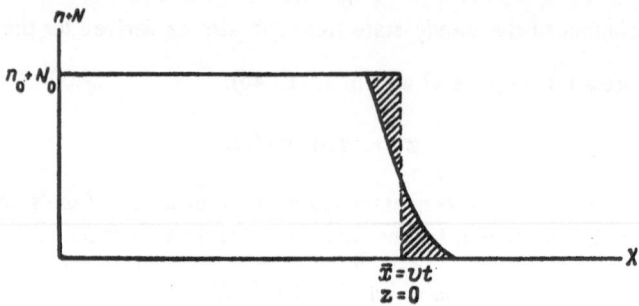

Fig. 10. For the theory of the stationary front of sorption dynamics in the case of a convex isotherm. Determination of the effective width of the zone and effective concentration in the front.

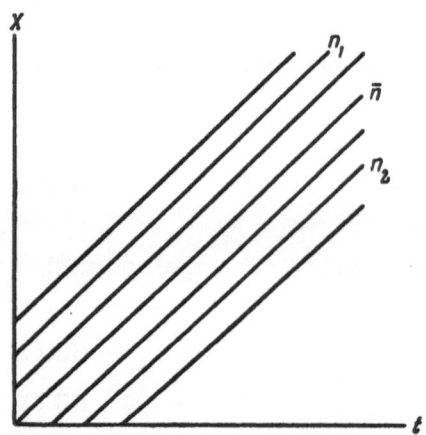

Fig. 11. On the theory of the stationary front of sorption dynamics with a convex isotherm. Family of asymptotes geometrically representing the equation of motion of the stationary front ($n_1 < \bar{n} < n_2$).

In the left-hand portion of this equation, the first member denotes the amount of the substance in the mobile phase, the second member that in the sorbent. Here we used integration with respect to the variables n and N, since the equations of the stationary front usually take the form of explicit functions with respect to the coordinate x. When $D^* \neq 0$, the coordinate x is defined by equations (III.45) and (III.59). Substituting the right-hand portions of these equations in place of x in the expressions under the integral sign in equation (III.65), we obtain

$$\int_0^{n_0} [vt + z(n)]\, dn$$

$$+ \int_0^{N_0} [vt + z(N)]\, dN = n_0 vt + N_0 vt. \qquad \text{(III.66)}$$

From the last, equation, after simplification, we shall have

$$\int_0^{n_0} z(n)\, dn + \int_0^{N_0} z(N)\, dN = 0. \qquad \text{(III.67)}$$

This expression permits us to give an unambiguous determination of the integration constant C in equations (III.58) and (III.62). Substituting (III.63) and (III.64) into (III.67), we obtain

$$C = - \frac{\int_0^{n_0} z_0(n)\, dn + \int_0^{N_0} z_0(N)\, dN}{n_0 + N_0}. \qquad \text{(III.68)}$$

Let us once again take up a number of properties of the asymptotic equation of the stationary frton (III.45). Geometrically it corresponds to a family of asymptotes, each of which gives the graph of the motion of a set concentration point n (Fig. 11). All the asymptotes in Fig. 11 possess the same slope. The slope of the straight lines is numerically equal to the rate of motion of the stationary front v. Among the family of asymptotes, there is one singular point $\bar{n}$, for which the function z(n) becomes zero. This singular, or, as we shall call it, effective concentration can be found directly by solving the equation $z(\bar{n}) = 0$. But, as was shown above, the condition z = 0 is the beginning of the reading in the new moving coordinate system when the variables are replaced by the substitution z = x−vt. The effective concentration $\bar{n}$ corresponds to the coordinate $\bar{x} = vt$ (see Fig. 10).

The method of obtaining the asymptotic solution of the problem·of sorption dynamics with a convex isotherm, considered here, was developed by Ya. B. Zel'dovich [63] and O. M. Todes [139, 140]. This method possesses a general value for the solution of the problems of sorption dynamics with a convex isotherm and the action of factors of blurring of the sorption front. Hence we shall henceforth call it the Zel'dovich-Todes method.

As was already noted above, long before the development of the theory of the stationary front of sorption dynamics, N. A. Shilov [168] proposed an empirical formula, called by his name. According to N. A. Shilov, the time of protective action of a sorption filter (moment of appearance of the substance to be sorbed at the outlet from the filter) is determined by the following formula:

$$t = Kx - \tau, \tag{III.69}$$

where K is the coefficient of the protective action; x is the length of the sorbent layer; τ is the loss of time of protective action. Let us rewrite equation (III.69) in a different form:

$$x = \frac{1}{K}t + \frac{\tau}{K}. \tag{III.70}$$

A comparison of equations (III.70) and (III.45) shows that the Shilov formula is a result of equation (III.45).

The coefficient of protective action is the reciprocal of the rate of motion of the stationary front, $K = 1/v$. The other quantity, τ/K, is the parameter $z(n)$ that we already know.

At a set length of the column of sorbent and a set concentration n_i indexed at the exit from the column, the quantity $z(n_i)$ takes a definite value. Thus,

$$\tau/K = \tau v = z(n_i),$$

from which

$$\tau = z(n_i)/v, \tag{III.71}$$

In honor of N. A. Shilov, who first established the formula of the motion of a stationary front empirically, we shall call equation (III.45) the Shilov equation.

The first elucidation of the conditions of formation of a stationary front of sorption dynamics and a theoretical interpretation of the Shilov empirical formula was given by Ya. B. Zel'dovich [63].

The Linear Isotherm. The linear sorption isotherm of one substance is usually written in the following form:

$$N = kn, \tag{III.72}$$

where k is the sorption coefficient or constant, sometimes called the Henry coefficient. The initial concentration of the substance introduced into the column n_0 will correspond to the equilibrium concentration of the substance in the sorbent N_0:

$$N_0 = kn_0. \tag{III.73}$$

Hence the sorption constant is numerically equal to the reciprocal of the partition ratio $k = 1/h = N_0/n_0$. On the basis of this, we shall henceforth write the linear sorption isotherm in the following form:

$$N = \frac{1}{h} n. \tag{III.74}$$

Substituting equation (III.74) into the equation of the balance (II.26), we obtain

$$\frac{\partial n}{\partial t} + u \frac{\partial n}{\partial x} + \frac{1}{h} \frac{\partial n}{\partial t} = D^\bullet \frac{\partial^2 n}{\partial x^2}, \tag{III.75}$$

from which we have after transformation

$$\frac{\partial n}{\partial t} + v\,\frac{\partial n}{\partial x} = H\,\frac{\partial^2 n}{\partial x^2}\,,\qquad\qquad\text{(III.76)}$$

where $v = uh/(1 + h)$; $H = D^* h/(1 + h)$.

Exact solutions of equation (III.76) for the initial and boundary conditions (III.46)-(III.48) are known in the literature [17, 182, 210]. However, these solutions represent complex mathematical expressions from series of special functions and are little suited to practice. In the theory of diffusion [17], there is an approximate asymptotic solution, which, with an accuracy sufficient for practical purposes, can describe the dynamics of equilibrium sorption in the case of a linear isotherm for the case $D^* \neq 0$.

The approximate asymptotic solution will be [235]:

$$\frac{n}{n_0} = \frac{N}{N_0} = 0.5\left[1 - \operatorname{erf}\left(\frac{x - vt}{2\sqrt{Ht}}\right)\right],\qquad\qquad\text{(III.77)}$$

where

$$\operatorname{erf}(w) = \frac{2}{\sqrt{\pi}}\int_0^w e^{-w^2}dw\qquad\qquad\text{(III.78)}$$

the so-called Kramp integral. The function erf(w) has been tabulated, and tables of it are given in many mathematical handbooks.

A comparison of the solution (III.77) with the exact solution shows [6] that the solution (III.77) describes the dynamic distribution curves in the interval of concentrations $\varphi = n/n_0$ and $\vartheta = N/N_0$ of approximately 0.1-0.9 with sufficient accuracy (error < 5%). When $\varphi < 0.1$ and $\varphi > 0.9$, the error in the asymptotic approximate solution becomes substantial. However, it is precisely in these concentration regions that the measurement errors also become substantial.

Let us consider a number of results of equation (III.77). Let us introduce the notations $n/n_0 = \varphi$ and $N/N_0 = \vartheta$. For a linear isotherm, $\varphi = \vartheta$. Let us write equation (III.77) in the following form:

$$\varphi = 0.5\,[1 - \operatorname{erf}(w)],\qquad\qquad\text{(III.79)}$$

where

$$w = \frac{x - vt}{2\sqrt{Ht}}\,.\qquad\qquad\text{(III.80)}$$

Then we have

$$\operatorname{erf}(w) = 1 - \frac{\varphi}{0.5}\,,\qquad\qquad\text{(III.81)}$$

i.e., both the function erf(w) itself and its argument w — a function of the concentration φ.

When $\varphi = 0.5$, the function erf(w) = 0 and $w_{0.5} = 0$, according to formula (III.80), we have $x_{0.5} - vt = 0$. This means that the point of the front $\varphi = 0.5$ should move along the column at constant speed:

$$v_{0.5} = x_{0.5}/t = v = u\,\frac{h}{1 + h} = \text{const},\qquad\qquad\text{(III.82)}$$

which discloses to us the physical meaning of the constant v in equation (III.76). In this equation, the quantity v determines the rate of motion of the point of half-concentration $\varphi = 0.5$. When $\varphi < 0.5$, the function erf(w) > 0, and w > 0. When $\varphi > 0.5$, the function erf(w) < 0 and w < 0.

However, if we take two concentration points in the region of concentrations $\varphi < 0.5$ and $\varphi > 0.5$, symmetrically situated with respect to $\varphi = 0.5$, namely $-\varphi_i$ and $1 - \varphi_i$, then on the basis of (III.81), we have:

44

$$| \operatorname{erf} (w_{\varphi_i}) | \; = \; | \operatorname{erf} (w_{1-\varphi_i}) | \quad \text{and} \quad | w_{\varphi_i} | \; = \; | w_{1-\varphi_i} | = w;$$

for $\varphi < 0.5$, $w = + |w|$; for $\varphi > 0.5$, $w = - |w|$.

The equation of motion of the sorption front can be written on the basis of (III.80) in the form

$$x = vt \pm 2w \sqrt{Ht}, \tag{III.83}$$

where the sign is positive for $\varphi < 0.5$ and negative for $\varphi > 0.5$. Let us determine the width of the sorption front in the following way:

$$\delta_x = x_{\varphi_i} - x_{1-\varphi_i}.$$

then from equation (III.83) we shall have

$$\delta_x = 4w \sqrt{Ht}, \tag{III.84}$$

i.e., the width of the sorption front increases in proportion to $\sqrt{t}$. This is one of the basic quasidiffusion principles of sorption dynamics for a linear isotherm.

From the analysis cited, it follows that only a single point of the front $\varphi = 0.5$ will move at a constant velocity $v_{0.5} = uh/(1+h)$; all other points of the front will move at the velocity

$$v_\varphi = \frac{dx}{dt} = v \pm w \sqrt{\frac{H}{t}}, \tag{III.85}$$

where for $\varphi < 0.5$, $v_\varphi > v$, while for $\varphi > 0.5$, $v_\varphi < v$, which is also a sign of increasing blurring of the front.

It is of interest to consider still another particular case — the dynamics of transport of a nonabsorbed substance in a column of a porous medium. In the absence of sorption, the value of the partition ratio $h \to \infty$ when ($N_0 = 0$); hence $v = u$, $H = D^*$, and the course of the dynamic distribution of the substance in the column is described by the equation

$$\varphi = \frac{n}{n_0} = 0.5 \left[1 - \operatorname{erf} \left(\frac{x - ut}{2 \sqrt{D^* t}} \right) \right]. \tag{III.86}$$

Then the results that have already been established earlier will be analogous. In formulas (III.80), (III.83), and (III.84), we need only replace v by u and H by D^*. Here also only one point $\varphi = 0.5$ will move at a constant velocity $u_{0.5} = u$, where u is the average rate of flow, and the width of the front will increase in proportion to $\sqrt{t}$.

The Concave Isotherm. Under the conditions of equilibrium sorption dynamics, the blurring of the front due to concavity of the isotherm is proportional to t, while the longitudinal effects of blurring ($D^* \neq 0$) lead to blurring of the front in proportion to $\sqrt{t}$. Thus, the first factor of blurring (concavity of the isotherm) is a stronger factor, and in a first approximation the longitudinal effects can be disregarded. Then the asymptotic distribution of the substance along the column of the sorbent can be approximately described using Wicke's law.

Hence, for a concave sorption isotherm under conditions of the action of longitudinal effects, the following pattern should be observed: the width of the front increases approximately in proportion to the time of the process. This case was not investigated in any greater detail.

4. Nonequilibrium Sorption Dynamics in the Absence of Longitudinal Effects ($D^* = 0$)

Sorption dynamics under nonequilibrium conditions also depend on the type of the sorption isotherm. The factor of the sorption kinetics is another factor in the blurring of the front. The same considerations that were outlined in the preceding section bring us to the following general picture of sorption dynamics under nonequi-

45

librium conditions: in the case of a convex sorption isotherm, a stationary front should be formed at the asymptotic stage of the process; in the case of a linear isotherm, an expanding front should be formed (the width of the front increases in proportion to $\sqrt{t}$); in the case of a concave isotherm, an expanding front should also be formed (width of the front increases in proportion to t).

The system of differential equations describing the nonequilibrium sorption dynamics when $D^* = 0$ will consist of the following equations:

$$\frac{\partial n}{\partial t} + u\,\frac{\partial n}{\partial x} + \frac{\partial N}{\partial t} = 0;$$ (III.5)

$$\frac{\partial N}{\partial t} = \psi\,(n,\,N).$$ (II.27)

Conditions (III.1), (III.3), and (III.4) should be taken as the initial and boundary conditions. For the asymptotic stage of sorption dynamics (when $t \to \infty$), equilibrium sets in asymptotically in the upper layers of the column: $N = f(t) \to N_0$. Hence the boundary conditions (III.3) and (III.4) should be replaced by asymptotic boundary conditions

$$t = \infty,\; x = 0,\; n = n_0,\; N = N_0;$$ (III.87)

$$x = \infty,\; n = 0,\; N = 0.$$ (III.88)

Let us consider the asymptotic solutions in general form as a function of the type of the isotherm.

The Convex Isotherm. Let us use the Zel'dovich-Todes method to solve the problem [6, 63, 140]. Let us introduce the variable (III.42), where v is the velocity of the stationary front. For the asymptotic stage, the boundary conditions (III.87) and (III.88) will take the following form when the new variable z is introduced:

$$t = \infty,\; z = -\infty,\; n = n_0,\; N = N_0;$$ (III.89)

$$z = +\infty,\; n = 0,\; N = 0.$$ (III.90)

Substitution of $z = x - vt$ provides the possibility of transforming equations (III.5) and (II.27) from equations in partial derivatives to the usual differential equations:

$$-v\,\frac{dn}{dt} + u\,\frac{dn}{dz} - v\,\frac{dN}{dz} = 0;$$ (III.91)

$$-v\,\frac{dN}{dz} = \psi\,(n,\,N).$$ (III.92)

After integration of the equation of the balance (III.91) with respect to z and determining the integration constant, on the basis of the boundary condition (III.90), we obtain

$$(u - v)\,n - vN = 0,$$ (III.93)

from which we find the rate of motion of the stationary front

$$v = u\,\frac{n}{n + N}.$$ (III.94)

But, on the other hand, substituting the concentration of the substance into (III.93), according to the boundary condition (III.89), we have

$$(u - v)\,n_0 - vN_0 = 0,$$ (III.95)

from which the same rate of motion of the stationary front will be

$$v = u\,\frac{n_0}{n_0 + N_0}.$$ (III.96)

Equality of the right-hand portions of formulas (III.94) and (III.96) gives an important function, characterizing the distribution of the substance between the mobile phase and the sorbent in a stationary front:

$$\frac{n}{n_0} = \frac{N}{N_0} \quad \text{or} \quad n = hN. \tag{III.97}$$

This means that in a stationary front of nonequilibrium sorption dynamics when $D^* = 0$, a linear relationship should be observed between the nonequilibrium concentrations of the substance in the mobile phase and in the sorbent. Although this ratio follows directly from the equation of the balance using the boundary conditions for the asymptotic stage of the process, however, in view of the absence of the factor of longitudinal transport in the equation ($D^* = 0$), it represents the characteristic law for a stationary front of nonequilibrium sorption dynamics when $D^* = 0$.

Function (III.97) was first established theoretically by Ya. B. Zel'dovich [63, 139]. Hence we shall call it the Zel'dovich function.

To obtain the equation of motion of the stationary front $f(n, x, t) = 0$, we should use the equation of the sorption kinetics (III.92). Substituting $n = hN$ and $v = uh/(1 + h)$ into this equation and integrating it, we obtain the asymptotic equation of the motion of a stationary front in general form:

$$z(N) = -\frac{uh}{1+h} \int \frac{dN}{\psi(N)} + C \tag{III.98}$$

or

$$z(n) = -\frac{1}{1+h} \int \frac{dn}{\psi(n)} + C, \tag{III.99}$$

where the integration constant C can be determined from the law of conservation of matter by means of equations (III.67) and (III.68), as was discussed above.

Hence, the nonequilibrium sorption dynamics in the case of a convex isotherm and $D^* = 0$ obey a general law at the asymptotic stage of the process—the Shilov equation (III.45) and a specific law—the Zel'dovich function (III.97).

The Linear Isotherm. Let us use the approximate equations of the method of "lagging coordinates" (II.32) or (II.33) to consider the problem of nonequilibrium sorption dynamics in the case of a linear isotherm. We resort to them because we are not considering any concrete mechanisms or types of the sorption process within the framework of the problem of the general theory of sorption dynamics. The equations of the sorption kinetics in the method of "lagging coordinates" [11-13, 163-165], as has already been noted earlier, consider the nonequilibrium character of the sorption process in generalized form, without considering the mechanisms of the kinetics, through the dynamic parameters of the lag path $\varkappa$ or the lag time τ.

In the case of a linear sorption isotherm, $N = f(n) = (1/h)n$. Hence, the two variations of the equation of sorption kinetics—(III.32) and (III.33)—take the following form in the method of lagging coordinates after substitution of the sorption isotherm (III.74):

$$\frac{\partial N}{\partial t} = \frac{1}{h} \cdot \frac{\partial n}{\partial t} + \varkappa \cdot \frac{1}{h} \cdot \frac{\partial^2 n}{\partial x \partial t} \tag{III.100}$$

or

$$\frac{\partial N}{\partial t} = \frac{1}{h} \frac{\partial n}{\partial t} - \tau \cdot \frac{1}{h} \cdot \frac{\partial^2 n}{\partial t^2}. \tag{III.101}$$

Substituting the right-hand portions of these kinetic equations into the equation of the balance (III.5), we obtain the following two forms of the initial differential equations of nonequilibrium sorption dynamics with a linear isotherm:

$$\frac{\partial n}{\partial t} + v\,\frac{\partial n}{\partial x} = -\,\frac{\varkappa}{1+h}\,\frac{\partial^2 n}{\partial x\partial t}\,, \qquad\qquad\qquad (\text{III.102})$$

$$\frac{\partial n}{\partial t} + v\,\frac{\partial n}{\partial x} = \frac{\tau}{1+h}\,\frac{\partial^2 n}{dt^2}\,, \qquad\qquad\qquad (\text{III.103})$$

where $v = uh/(1+h)$.

Solutions of equation (III.102) or (III.103) can be obtained by operational calculus. Within definite limits $\varphi = 0.1\text{-}0.9$, these solutions can be approximated by the following approximate equivalent equations [154, 163, 164].

$$\frac{N}{N_0} \approx \frac{n}{n_0} = 0.5\left[1 - \operatorname{erf}\left(\frac{x - vt}{2\sqrt{\dfrac{\varkappa v}{1+h}\cdot t}}\right)\right]; \qquad\qquad (\text{III.104})$$

$$\frac{N}{N_0} \approx \frac{n}{n_0} = 0.5\left[1 - \operatorname{erf}\left(\frac{x - vt}{2\sqrt{\dfrac{\tau v^2}{1+h}\cdot t}}\right)\right]. \qquad\qquad (\text{III.105})$$

We have already considered an analogous solution [see (III.77)]. Just as before, in a consideration of the properties of equation (III.77), in this case also it can be shown that there will be only one point that moves with a constant velocity among the points of the front—the point of half-concentration $\varphi = 0.5$:

$$v_{0.5} = v = u\,\frac{h}{1+h}.$$

The front of the substance to be sorbed will expand in proportion to $\sqrt{t}$.

A comparison of equations (III.104) and (III.105) gives the following relationship between the dynamic constants $\varkappa$ and τ:

$$\frac{\varkappa}{\tau} = v_{0.5} = u\,\frac{h}{1+h}. \qquad\qquad\qquad (\text{III.106})$$

On the other hand, equations (III.104) and (III.105) can be used to give a quasidiffusion treatment on the basis of a comparison of them with equation (III.77), which actually differs from (III.104) and (III.105) only in the symbols of the constants. On the basis of this comparison, let us introduce the concept of the quasidiffusion kinetic constant of blurring D_q.

The relationship of this constant to the dynamic constants $\varkappa$ and τ will be the following:

$$D_q^* = \frac{\varkappa u}{1+h} = \frac{h}{(1+h)^3}\,u^2\tau. \qquad\qquad\qquad (\text{III.107})$$

The Concave Isotherm. In the case of a concave isotherm, in a first approximation we can neglect the kinetic factor of blurring and consider the asymptotic stage of sorption dynamics within the framework of the theory of equilibrium sorption dynamics with a concave isotherm, using Wicke's law. This means that in the process of dynamics there will be a progressive blurring of the front. Moreover, the width of the front will increase approximately in proportion to t.

5. Nonequilibrium Sorption Dynamics under the Action of Longitudinal Effects $(D^* \neq 0)$

The initial differential equations for this problem are equations (II.26) and (II.27). This is already a more complex problem. In the preceding sections, we obtained relatively simple solutions, thanks to the introduction of a number of simplifying assumptions. The system of equations (II.26) and (II.27), which describes the problem

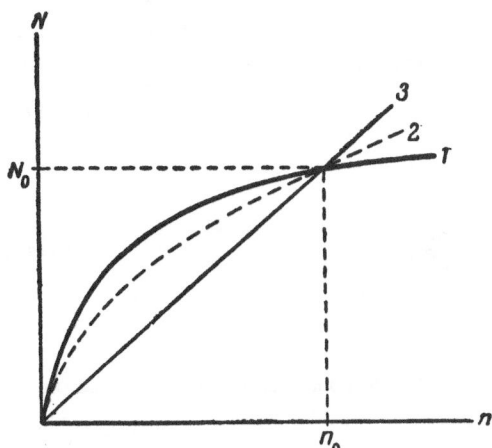

Fig. 12. Ratio between the concentrations of the substance in the sorbent and in the mobile phase in the stationary front. 1) Equilibrium sorption dynamics under the action of longitudinal effects (sorption isotherm); 2) nonequilibrium sorption dynamics under the action of longitudinal effects [equation (III.108)]; 3) nonequilibrium sorption dynamics in the absence of longitudinal effects (linear function).

of nonequilibrium sorption dynamics under the action of longitudinal effects ($D^* \neq 0$), more fully reflects the real process of sorption dynamics in the case when all factors of blurring operate simultaneously: kinetic, diffusion, hydrodynamic, etc.

No solution of the problem posed in general form has yet been found. Hence, let us outline only certain approaches to its solution.

The Convex Isotherm. Since the kinetic and other factors of blurring operate in the same direction—they blur the sorption front—then the factor of convexity of the isotherm should lead to the formation of a stationary front at the asymptotic stage of the process [140]. For the asymptotic stage of the stationary front, we have the already considered boundary conditions (III.87), (III.88).

Introducing the substitution $z = x - vt$, where v is the velocity of the stationary front, we obtain from equation (II.26) the equation (III.54), and from (II.27), the equation (III.92).

In this case, the rate of motion of the stationary front, as was shown earlier, will also be determined by the already known formula

$$v = _{,} u \, \frac{h}{1 + h}.$$

But in contrast to the conditions of equilibrium sorption dynamics when $D^* \neq 0$ and the conditions of nonequilibrium sorption dynamics when $D^* = 0$, the ratio between the concentrations n and N in the front will obey a more complex dependence. This dependence can be obtained from equations (III.54) and (III.92) by eliminating the variable from them:

$$\frac{dN}{dn} = - \frac{D^*}{u^2} \cdot \frac{(1 + h)^2}{h} \cdot \frac{\psi(n, N)}{n - hN}. \tag{III.108}$$

Unfortunately, this differential equation is an equation with unresolved variables, and special approximate methods must be used for its integration [98]. But for this we need to know the concrete form of the kinetic function $\psi(n, N)$. If we can find the ratio between the nonequilibrium concentrations in the front $N = f_{fr}(n)$ for a concrete form of the function $\psi(n, N)$, then, by substituting this ratio into the equation (III.92), we can obtain the equation of the stationary front characterizing the distribution of the substance to be sorbed along the column.

As it follows from (III.108), the ratio between the nonequilibrium concentrations n and N in the front will possess some intermediate character between the equilibrium distribution and the distribution according to the linear Zel'dovich function. This means that the function $N = f_{fr}(n)$ will be found somewhere in the region between the two functions: the function of the convex sorption isotherm and the linear Zel'dovich function (Fig. 12).

The Linear Isotherm. As was already indicated above, in the case of a linear isotherm, by using the semiphenomenological method of lagging coordinates, we can consider the kinetic blurring of the front as a quasidiffusion process, characterized by the quasidiffusion constant D_q^*. Then the corresponding differential equation of the nonequilibrium sorption dynamics can be written in the form

$$\frac{\partial n}{\partial t} + v \, \frac{\partial n}{\partial x} = D_q^* \, \frac{h}{1 + h} \cdot \frac{\partial^2 n}{\partial x^2}. \tag{III.109}$$

49

In the problem under consideration, in addition to the sorption kinetics, which can be formally characterized by the constant D_q^*, longitudinal effects also operate. For them let us introduce the effective constant of longitudinal effects D_l^*. The resultant quasidiffusion constant can be considered as a sum of the two constants

$$D^* = D_q^* + D_l^* \qquad \qquad \text{(III.110)}$$

and the general equation will be

$$\frac{\partial n}{\partial t} + v\,\frac{\partial n}{\partial x} = D^*\,\frac{h}{1+h}\cdot\frac{\partial^2 n}{\partial x^2}\,, \qquad \qquad \text{(III.111)}$$

We already know the solution for this equation [see, for example, (III.77)].

In concluding this section, let us note that the problem of nonequilibrium sorption dynamics under the action of longitudinal effects will require a further theoretical investigation.

CHAPTER IV

THEORY OF FRONTAL CHROMATOGRAPHY

1. Equilibrium Sorption Dynamics of a Mixture of Substances in the Absence of Longitudinal Effects ($D^* = 0$)

Depending on the sorbability of the components of the mixture on a given sorbent, three limiting cases of sorption dynamics can be encountered: 1) all the components of the mixture of substances possess the same sorbability; 2) all the components of the mixture possess different sorbabilities, but they are sorbed according to the linear isotherm type (independent sorption of each component); 3) all the components of the mixture possess different sorbabilities, but they are absorbed according to the type of mutually dependent sorption (the sorption of each component depends on the concentration of the other components of the mixture).

In the first case, a mixture of identically sorbed substances can be considered as one substance, and the problem of sorption dynamics of such a mixture reduces to the problem of the sorption dynamics of one substance, of which we spoke in the preceding chapter.

In the second case, since each component of the mixture is sorbed independent of the presence of the other components, the problem of sorption dynamics of the mixture of substances also reduces to a problem of sorption dynamics of one substance.

The third case is the most complex. It also represents the main object of study in the theory of sorption dynamics of a mixture of substances and chromatography. To describe the process of sorption dynamics of a mixture of substances, let us consider the following system of equations:

$$\frac{\partial n_i}{\partial t} + u \frac{\partial n_i}{\partial x} + \frac{\partial N_i}{\partial t} = D^* \frac{\partial^2 n_i}{\partial x^2}, \tag{II.21}$$

$$\frac{\partial N_i}{\partial t} = \psi_i (n_1, n_2, \ldots, n_j, N_1, N_2, \ldots, N_j),$$
$$1 \leqslant i \leqslant j. \tag{II.4'}$$

Moreover, let us assume that in all the components of the mixture the sorbability is different and interdependent. Just as before, to obtain some sort of results of practical importance, let us take the road of simplifying the problem. Hence, the first step in the theory will be a consideration of the basic principles of equilibrium sorption dynamics of a mixture of substances.

Let us propose first that the equilibrium sorption dynamics is accomplished in the absence of longitudinal effects of blurring ($D^* = 0$). Then the initial system of differential equations will take the following form:

$$\frac{\partial n_i}{\partial t} + u \frac{\partial n_i}{\partial x} + \frac{\partial N_i}{\partial t} = 0; \tag{IV.1}$$

$$N_i = f_i (n_1, n_2, \ldots, n_j),$$
$$1 \leqslant i \leqslant j. \tag{II.6}$$

Let the components of the mixture comprise the following sorption series:

$$A_1 < A_2 < \ldots < A_j. \tag{IV.2}$$

Now let us set definite initial and boundary conditions for the frontal sorption dynamics of a mixture of substances. Let us assume that at the initial moment there is already some initial equilibrium distribution of each of the components of the mixture in the column:

$$t = 0, \quad 0 \leqslant x \leqslant x_{0,i}, \quad n_i = \varphi_i(x), \quad N_i = f_i(n_1, n_2, \ldots, n_j), \qquad \text{(IV.3)}$$

$$x > x_{0,i}, \quad n_i = 0, \quad N_i = 0,$$

$$1 \leqslant i \leqslant j,$$

where $\varphi_i(x)$ represent set continuous differentiable functions of the initial distribution of the components. To avoid creating a discontinuity of the distribution function at the entrance to the column, let us set the following boundary conditions:

$$x = 0, \quad t > 0, \quad n_i = n_i^0 = \varphi_i(0);$$

$$N_i = N_i^0 = f_i(n_1^0, n_2^0, \ldots, n_j^0); \qquad \text{(IV.4)}$$

$$x = \infty, \quad t > 0, \quad n_i = 0, \quad N_i = 0. \qquad \text{(IV.5)}$$

This means that at the entrance to the column, the mobile phase is constantly delivered with the initial concentrations of the components $n_1^0, n_2^0, \ldots, n_j^0$, while as a result of the equilibrium of the process, equilibrium concentrations in the sorbent $N_1^0, \ldots, N_j^0$ are instantaneously established and remain constant at the entrance to the column at the initial moment and all subsequent moments, in accord with the sorption isotherms (II.6).

Let us eliminate the variables N_i from equation (IV.1) by replacement of $\partial N_i / \partial t$ by the derivative of the isotherm (II.6):

$$\frac{\partial N_i}{\partial t} = \frac{\partial f_i}{\partial n_1} \cdot \frac{\partial n_1}{\partial t} + \frac{\partial f_i}{\partial n_2} \cdot \frac{\partial n_2}{\partial t} + \cdots + \frac{\partial f_i}{\partial n_j} \cdot \frac{\partial n_j}{\partial t}. \qquad \text{(IV.6)}$$

After substitution of (IV.6) into (IV.1), we obtain

$$\left.\begin{aligned}
&\frac{\partial n_1}{\partial t} + u\frac{\partial n_1}{\partial x} + f'_{1,1}\frac{\partial n_1}{\partial t} + f'_{1,2}\frac{\partial n_2}{\partial t} + \cdots + f'_{1,j}\frac{\partial n_j}{\partial t} = 0;\\
&\frac{\partial n_2}{\partial t} + u\frac{\partial n_2}{\partial x} + f'_{2,1}\frac{\partial n_1}{\partial t} + f'_{2,2}\frac{\partial n_2}{\partial t} + \cdots + f'_{2,j}\frac{\partial n_j}{\partial t} = 0;\\
&\cdots\cdots\cdots\cdots\cdots\cdots\cdots\cdots\cdots\cdots\cdots\\
&\frac{\partial n_j}{\partial t} + u\frac{\partial n_j}{\partial x} + f'_{j,1}\frac{\partial n_1}{\partial t} + f'_{j,2}\frac{\partial n_2}{\partial t} + \cdots + f'_{j,j}\frac{\partial n_j}{\partial t} = 0.
\end{aligned}\right\} \qquad \text{(IV.7)}$$

Then, grouping like factors;

$$\left.\begin{aligned}
&u\frac{\partial n_1}{\partial x} + (1 + f'_{1,1})\frac{\partial n_1}{\partial t} + f'_{1,2}\frac{\partial n_2}{\partial t} + \cdots + f'_{1,j}\frac{\partial n_j}{\partial t} = 0;\\
&f'_{2,1}\frac{\partial n_1}{\partial t} + u\frac{\partial n_2}{\partial x} + (1 + f'_{2,2})\frac{\partial n_2}{\partial t} + \cdots + f'_{2,j}\frac{\partial n_j}{\partial t} = 0;\\
&\cdots\cdots\cdots\cdots\cdots\cdots\cdots\cdots\cdots\cdots\cdots\\
&f'_{j,1}\frac{\partial n_1}{\partial t} + f'_{j,2}\frac{\partial n_2}{\partial t} + \cdots + u\frac{\partial n_j}{\partial x} + (1 + f'_{j,j})\frac{\partial n_j}{\partial t} = 0.
\end{aligned}\right\} \qquad \text{(IV.8)}$$

We shall seek the solution of this system by the method of replacing variables, analogously to our procedure in the solution of the problem of equilibrium sorption dynamics of one substance. Then we need to select a new variable, with the aid of which we might transform the system of equations in partial derivatives to a system of the usual differential equations. We shall seek a solution of the system in the form of certain functions of a new parameter $z = x - vt$, where v is some quantity possessing the physical meaning of the rate of motion of the concentration points. We as yet can say nothing with respect to the character of this value of the velocity—whether it is general or some sort of set of concentration points. The answer to this question should give the very solution of the problem posed.

Hence, let us assume that the solution of the system (IV.8) should take the following form:

$$\left.\begin{array}{l} n_1 = \varphi_1\,(x - vt) = \varphi_1\,(z); \\ n_2 = \varphi_2\,(x - vt) = \varphi_2\,(z); \\ \cdots\cdots\cdots\cdots\cdots \\ n_i = \varphi_j\,(x - vt) = \varphi_j\,(z). \end{array}\right\} \tag{IV.9}$$

Using this proposed solution, let us replace the variables in the system (IV.8). After the proper tranformations, we obtain

$$\left.\begin{array}{l} (f'_{1,1} - \lambda)\,\dfrac{d\varphi_1}{dz} + f'_{1,2}\,\dfrac{d\varphi_2}{dz} + \ldots + f'_{1,j}\,\dfrac{d\varphi_j}{dz} = 0; \\[2mm] f'_{2,1}\,\dfrac{d\varphi_1}{dz} + (f'_{2,2} - \lambda)\,\dfrac{d\varphi_2}{dz} + \ldots + f'_{2,j}\,\dfrac{d\varphi_j}{dz} = 0; \\[2mm] \cdots\cdots\cdots\cdots\cdots\cdots\cdots\cdots\cdots \\[2mm] f'_{j,1}\,\dfrac{d\varphi_1}{dz} + f'_{j,2}\,\dfrac{d\varphi_2}{dz} + \ldots + (f'_{j,j} - \lambda)\,\dfrac{d\varphi_j}{dz} = 0, \end{array}\right\} \tag{IV.10}$$

where $\lambda = (u/v) - 1$, $f'_{i,k} = \partial N_i / \partial n_k$ are partial derivatives of the sorption isotherms (II.6).

The system of equations obtained belongs to the type of systems of homogeneous linear equations. It possesses a trivial "zero" solution:

$$\frac{d\varphi_1}{dz} = 0; \quad \frac{d\varphi_2}{dz} = 0; \ldots; \quad \frac{d\varphi_j}{dz} = 0, \tag{IV.11}$$

from which

$$\varphi_1 = \text{const}, \quad \varphi_2 = \text{const}, \ldots, \varphi_j = \text{const} \tag{IV.12}$$

or

$$n_1 = \text{const}, \; n_2 = \text{const}, \ldots, n_j = \text{const}. \tag{IV.13}$$

Just as in the case of the sorption dynamics of one substance, the solution of (IV.13) corresponds in physical meaning to the case of the motion of a single concentration point: if there is a distribution of the substance characterized by the concentration point of each component, then these single concentrations of the components remain constant. Other, nonzero solutions of the system (IV.10) will exist only in the case when the determinant of the system $D = 0$:

$$\begin{vmatrix} f'_{1,1} - \lambda & \ldots & \ldots & f'_{1,j} \\ \cdots & \cdots & \cdots & \cdots \\ \ldots & f'_{i,i} - \lambda & \ldots & \ldots \\ \cdots & \cdots & \cdots & \cdots \\ f'_{j,1} & \ldots & \ldots & f'_{j,j} - \lambda \end{vmatrix} = 0. \tag{IV.14}$$

The algebraic equation (IV.14) obtained is called characteristic. Let us consider a number of physical results that follow from equation (IV.14).

The parameter λ, according to the notation adopted in equations (IV.10), determines the rate of motion of the concentration points in the process of sorption dynamics:

$$v = \frac{u}{1 + \lambda}. \tag{IV.15}$$

But equation (IV.14) is an algebraic equation of the j-th order with respect to the unknown quantity λ, and, consequently, should possess j roots: $\lambda_1, \lambda_2, \ldots, \lambda_j$. From this we obtain the fact that the number of velocities of the concentration points will also be equal to j:

$$v_1 = \frac{u}{1 + \lambda_1}, \; v_2 = \frac{u}{1 + \lambda_2}, \; \ldots, \; v_j = \frac{u}{1 + \lambda_j}, \tag{IV.16}$$

i.e., the number of theoretically predicted velocities is equal to the number of components of the mixture of substances. Since equation (IV.14) contains derivatives of the sorption isotherms, then the characteristic parameters $\lambda_1, \ldots, \lambda_j$ are functions of the concentration:

$$\lambda_1 = \lambda_1(n_1, \ldots, n_j); \ldots; \lambda_j = \lambda_j(n_1, \ldots, n_j). \qquad \text{(IV.17)}$$

From this it follows that to a set aggregate of substances correspond j different velocities $v_1, \ldots, v_j$, each of which, according to (IV.16) and (IV.17), is a function of the concentration of all the components:

$$v_1 = v_1(n_1, n_2, \ldots, n_j); \ldots; v_j = v_j(n_1, \ldots, n_j). \qquad \text{(IV.18)}$$

We shall consider that all the roots of equation (IV.14) are real (only the real roots possess physical meaning). Since $\lambda = u/v - 1$, while $v \leq u$ (the velocity of a substance cannot exceed the flow rate), then λ is always greater than or equal to zero, i.e., the roots of equation (IV.14) should lie in the region of positive numerical values.

Let us determine the physical meaning of the theoretical prediction of multiple values of the rates of motion of the concentrations in the process of sorption dynamics of a mixture of substances.

First of all, let us note that the characteristic parameters acquire definite numerical values for each set, fixed aggregate of concentrations, as it follows from equation (IV.17). Let us explain this more graphically. At some moment of time, let some distribution of j substances be made up in the sorption column. Let us arbitrarily single out some portion of the column, as shown in Fig. 13. Let us take a definite coordinate x on this portion. At a given moment of time t, a definite, fixed set of concentration points $n_1, n_2, \ldots, n_j$ corresponds to this coordinate. For a given set, according to formula (IV.17), obtained as a result of solution of the characteristic equation (IV.10), we can calculate the numerical values of the characteristic parameters $\lambda_1, \ldots, \lambda_j$; and further—according to formulas (IV.16)—the numerical values of the velocities $v_1, \ldots, v_j$. What are these velocities?

A single physical interpretation of this set of velocities can be given: each concentration point of the front of any substance possesses its own characteristic velocity at each given moment; different concentration points from a given set $n_1, \ldots, n_j$ possess different velocities. Physically this should be so. Now, according to the conditions of the problem, the substances of the mixture possess different sorbabilities. It is precisely the different sorbabilities of the components of the mixture of substances that make up the first cause of the difference in the velocities and distributions of the substances in the sorption column during the process of sorption dynamics. This physical factor also gives the chromatographic zone distribution of substances and their separation. It is natural that substances with greater sorbability will move along the column more slowly, while substances with lower sorbability will move more rapidly. In the case of interdependent sorption of a mixture of substances, the different sorbabilities are responsible for the phenomenon of sorption displacement—a substance with greater sorbability will displace and push ahead of a substance with lower sorbability. The spectrum of velocities $v_1, v_2, \ldots, v_j$ obtained is easily separated according to affiliation among the corresponding components of the mixture. If the velocities possess numerical values such that $v_1 > v_2 > \ldots > v_j$, then the velocity v_1 will belong to the first member of the sorption series (IV.2), the velocity v_2 to the second member, etc. Thus, the relative sorbabilities of the components of a given mixture of substances can be judged by the value of the velocity.

Then, a serious complication is encountered along the way to solving the problem posed. Although each concentration point from a fixed set of concentrations $n_1, n_2, \ldots, n_j$ possesses its own definite characteristic velocity, however, this velocity is instantaneous, i.e., the velocity at a given moment of time t in a given coordinate x.

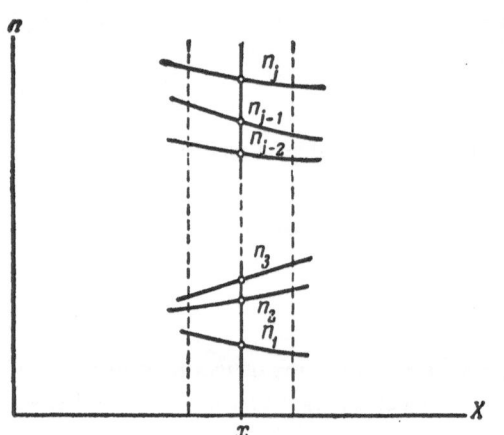

Fig. 13. For the theory of sorption dynamics of a mixture of substances. Set of concentration points of a mixture of substances in a set coordinate of the column.

Actually, since the velocities of the concentrations $n_1, n_2, \ldots, \dot{n}_j$, registered in Fig. 13, differ, then they will be displaced different distances along the X-axis. This leads to the fact that in the process of equilibrium sorption dynamics, a continuous instantaneous change in the ratios of the concentrations in each coordinate x will occur. In other words, if we fix some concentration, for example, n_i^*, and follow its motion, then we note that the concentrations of other substances corresponding to it vary continuously. But the variation of the ratio of the concentrations at each given moment t and in each coordinate x will be accompanied by a variation of the characteristic parameters $\lambda_1, \lambda_2, \ldots, \lambda_j$, and, consequently, by a variation of the velocities of the concentration points of the fronts as well.

Thus, Wicke's law will no longer be observed in the process of sorption dynamics of a mixture of substances. Although in the case of sorption dynamics of one substance, according to Wicke's law, each concentration point moves with its own characteristic, but constant velocity, in the case of the dynamics of interrelated sorption of a mixture of substances, each concentration point moves with its own characteristically varying (variable) velocity.

The entire complexity of the problem lies in the fact that the course of the time variation of the concentrations of the substances that accompany each set concentration must be established on the basis of the initial and boundary conditions. For example, if we follow the motion of some fixed concentration point n_i^* for the i-th substance, then we must know how the concentrations of all the other substances accompanying it vary with time during the process of sorption dynamics: $n_1(t), n_2(t), \ldots, n_{i-1}(t), n_{i+1}(t), \ldots, n_j(t)$. If we know these functions, in principle we can theoretically find the course of the time variation of the parameter $\lambda_i [n_1(t), \ldots, n_i^*, \ldots, n_j(t)]$, and then the dependence of the velocity on the time as well:

$$v_i = \frac{u}{1 + \lambda [n_1(t), \ldots, n_i^*, \ldots, n_j(t)]} . \qquad (\text{IV.19})$$

If we were to find the function $\lambda_i(t)$ or $v_i(t)$, then it would not be difficult to construct the solution of the problem of frontal equilibrium sorption dynamics posed. For example, at the initial moment, the concentration point n_i for the i-th substance, according to the initial conditions (IV.3), possessed the coordinate $x = \bar{\varphi}_i(n_i^*)$, where $\bar{\varphi}_i(n_i^*)$ is the reciprocal of $\varphi_i(x)$. The coordinate of this point at any moment of time t can be expressed by the following integral equation:

$$x = \bar{\varphi}_i(n_i^*) + \int_0^t v_i [n_1(t), \ldots, n_i^*, \ldots, n_j(t)]\, dt, \qquad (\text{IV.20})$$

where n_i^* is a set fixed concentration of the i-th component, independent of the time; the remaining concentrations are variable, dependent on the time. But the entire trouble lies in the fact that these time dependences of the concentrations cannot be found. This problem has not been resolved. It is doubtful that such a problem could be solved in general analytical form. In any case, further research is required here. One of the probable lines for further investigation is the use of methods of successive approximations. Undoubtedly the problem may be solved by the method of calculus of finite differences, but this is no longer a generalizing mathematical method. For its use, the conditions of the problem must be placed in as concrete terms as possible, setting concrete functions of the initial distribution of the substances, functions of the sorption isotherms, and other data.

The most general principle of equilibrium sorption dynamics of a mixture of substances, just as in the case of the sorption dynamics of one substance, is represented by the formulas of the velocities of the concentration points [(IV.15), (IV.16)], in conjunction with the characteristic equations (IV.14), defining the parameters λ.

Let us once again take up the role of the characteristic equation (IV.14) in the solution of theoretical problems of the sorption dynamics of a mixture of substances.

The characteristic equation (IV.14) expresses the influence on the value of the velocity of the concentration points not only of the concentrations of the component, but also of the form of the sorption isotherm, since equation (IV.14) contains partial derivatives of the sorption isotherms. Of course, the role of the type of isotherm in the case of the sorption dynamics of a mixture of substances is of a more complex character. Here two interrelated factors should exert an influence: 1) the form of the isotherm will predetermine the sorbabilities of the components of the mixture, and hence the course of their mutual sorption displacement during the process of sorption dynamics; 2) the character of the motion of various portions of the sorption fronts will depend on the shape of the isotherm. More briefly, the deformation of the distribution curves of the fronts of the substances in the process of sorption dynamics will depend on the sorbability of the components and on the concentration ratios.

By analogy with the theory of sorption dynamics of one substance, let us introduce the concept of "convexity" and "concavity" of the sorption isotherms, although in the case of polycomponent systems (multidimensional space), these concepts will possess a somewhat formal character.

We shall consider the isotherms "convex" if the values of their partial derivatives $\ldots, f'_{i,k}, \ldots$ decrease with increasing concentration of the substances, and, on the contrary, we shall consider them "concave" if the values of the partial derivatives increase with increasing concentration. Of course, in principle there can be more complex cases of dependence of the course of the variation of various partial derivatives on the concentrations of the components, i.e., there can be isotherms of a mixed type.

For the case of convex isotherms, qualitatively the same picture of deformation of the fronts should be observed as in the case of the sorption dynamics of one substance: smaller numerical values of the partial derivatives of the isotherms correspond to greater concentrations. In the limit, when $n_1, \ldots, n_j \rightarrow \infty$, the partial derivatives $f'_{i,k} \rightarrow 0$ according to (IV.14); also $\lambda \rightarrow 0$; correspondingly, according to formula (IV.15), we will have $v \rightarrow u$. Greater values of the derivatives and correspondingly greater values of λ should correspond to smaller concentrations; consequently, the velocities will also decrease. Thus, in the case of sorption dynamics of mixtures of substances as well, convexity of the sorption isotherms is a factor of compression and stabilization of the fronts of sorption dynamics.

For the case of concave isotherms, obviously the picture will be the opposite: concavity of the isotherm leads to a blurring of the fronts. Let us emphasize once again that the general theory of sorption dynamics of a mixture has been only touched upon by the investigation. Still more efforts should be applied to the establishment of the quantitative principles and the criteria of sorption dynamics of mixtures. Much still remains unclear. In particular, for example, a strictly quantitative establishment of the criteria of sorbability of the components of a mixture in blurred sorption dynamic fronts is essential.

M. S. Tsvet has experimentally established one of the qualitative laws of the sorption dynamics of mixtures of substances—the law of sorption displacement. Unfortunately, at present we cannot give any quantitative theoretical substantiation of this law, although we are confident that further development of the theory will make it possible to find a quantitative formulation of the law of sorption displacement in generalized form.

2. The Frontal Chromatogram

The frontal, or primary chromatogram is obtained in the process of continuous filtration of the initial mixture of substances through the sorption column.

Let us consider the simplest theory of the frontal chromatogram—the theory of equilibrium frontal sorption dynamics of a mixture in the absence of longitudinal effects of blurring ($D^* = 0$) [77, 241].

In contrast to the initial and boundary conditions used in the preceding section, here let us take conditions under which filtration of a mixture of substances passes through an originally "clean" sorption column:

$$t = 0, \ x > 0, \ n_i = 0, \ N_i = 0, \tag{IV.21}$$

$$x = 0, \ t \geqslant 0, \ n_i = n_i^0, \ N_i = N_i^0, \tag{IV.22}$$

$$x = \infty, t > 0, \ n_i = 0, \ N_i = 0, \tag{IV.23}$$
$$1 \leqslant i \leqslant j,$$

where n_i^0 is the concentration of the i-th component in the initial mixture of substances, N_i^0 is the equilibrium concentration of the i-th substance in the sorbent.

Let the basic condition of chromatography be accomplished—the substances of the mixture possess different sorbabilities and form a sorption series (IV.2). The problem consists of theoretically predicting the course of the distribution and motion of the substances in the frontal chromatogram.

In Chapter II, when we considered the sorption dynamics of one substance, it was shown that if there is a single concentration point at the entrance to the column at the initial moment, then it will move at a constant velocity $v = un_0/n_0 + N_0$ during the process of sorption dynamics, where n_0 and N_0 are invariant equilibrium concentrations in the mobile phase and in the sorbent. This phenomenon was substantiated as the physical interpretation of the "zero" solution of the equations of sorption dynamics. An analogous "zero" solution of the equations of sorption dynamics is also obtained in the case of sorption dynamics of a mixture, as was shown in the preceding section. This means that if in a sorption column each component of the mixture possesses an equilibrium concentration unique for it, and no factors create changes in the equilibrium conditions, then in the process of sorption dynamics, these concentrations should be maintained and should move at a constant velocity; the latter in turn should lead to the formulation of stable, stationary fronts with a straight break.

Let us use these considerations to elucidate the problem of how a frontal chromatogram arises under the initial and boundary conditions set above (IV.21)-(IV.23). First of all, let us attempt to give a quantitative criterion for the condition of formation of a frontal chromatogram.

In view of the fact that we are considering equilibrium sorption dynamics of a mixture, the process of emergence of a frontal chromatogram should be instantaneous, since all the equilibrium distributions in the column will be established instantaneously. If we take the beginning of the process, then at the initial moment, a series of events of distribution of substances at the entrance to the sorption column is simultaneously, instantaneously set up.

The first instantaneous process is the establishment of the equilibrium concentrations of the substances in the sorbent, corresponding to the initial concentrations of the substances $n_1^0, \ldots, n_j^0$, as was noted in the boundary condition (IV.22). Since for each substance there is a single concentration $n_1^0, \ldots, n_j^0$, and, correspondingly, a single concentration $N_1^0, \ldots, N_j^0$, the rates of motion at the initial moment should be determined by formulas analogous to the formulas of the rate (III.26) or (III.32), which is easily established on the basis of the law of conservation of matter.

Hence, the initial velocities of the concentrations set at the entrance should be the following:

$$v_1^0 = u \, \frac{n_1^0}{n_1^0 + N_1^0}, \quad v_2^0 = u \, \frac{n_2^0}{n_2^0 + N_2^0}, \quad \ldots, \quad v_j^0 = u \, \frac{n_j^0}{n_j^0 + N_j^0}. \tag{IV.24}$$

Introducing the concept of the distribution ratios

$$h_1^0 = \frac{n_1^0}{N_1^0}, \quad \ldots, \quad h_j^0 = \frac{n_j^0}{N_j^0}, \tag{IV.25}$$

we obtain

$$v_1^0 = u \, \frac{h_1^0}{1 + h_1^0}, \quad \ldots, \quad v_j^0 = u \, \frac{h_j^0}{1 + h_j^0}. \tag{IV.26}$$

Let us assume that the components are distributed between the mobile phase and the sorbent in such a way that

$$\frac{n_1^0}{N_1^0} = \ldots = \frac{n_j^0}{N_j^0} = \text{const}, \tag{IV.27}$$

i.e., in the same proportion, and, consequently, their distribution ratios are the same:

$$h_1^0 = \ldots = h_j^0 = \text{const}. \tag{IV.28}$$

In this case, the initial velocities of the concentrations set at the entrance should be the same, according to (IV.24) or (IV.26):

$$v_1^0 = \ldots = v_j^0 = \text{const.} \tag{IV.29}$$

The velocity of the concentration points of all the substances is constant, i.e., a straight stationary front of a mixed zone will move in the column at a constant velocity. No zonal distribution of the substances along the column occurs—one mixed zone of all the substances is formed. Thus, equality of the distribution ratios of the components of the mixture (IV.28) can be considered as a quantitative criterion of identical sorbability of the substances and the impossibility of formation of a frontal chromatogram.

In contrast to this, inequality of the distribution ratios (IV.28)

$$h_1^0 \neq \ldots \neq h_j^0 \neq \text{const} \tag{IV.30}$$

is the criterion for different sorbabilities of the substances. If this criterion is realized, then the initial velocities will differ:

$$v_1^0 \neq v_2^0 \neq \ldots \neq v_j^0 \neq \text{const.} \tag{IV.31}$$

Moreover, the smaller the distribution ratio h_i^0 for a given component, the lower its velocity v_i^0. The value of the distribution ratio in equilibrium sorption dynamics is the criterion for sorbability of a substance. The smaller the value of this ratio for a component of a mixture, the greater its sorbability. Since we numbered the components of the mixture in order of increasing sorbability [see the sorption series (IV.2)], we shall consider that

$$h_1^0 > h_2^0 > \ldots > h_j^0 \tag{IV.32}$$

and, correspondingly

$$v_1^0 > v_2^0 > \ldots > v_j^0. \tag{IV.33}$$

In the sorption series (IV.2), the j-th component possesses the greatest sorbability, while the first component possesses the least sorbability. Correspondingly, according to (IV.32) and (IV.33), the j-th component will possess the smallest distribution ratio h_j^0 and the lowest velocity v_j^0, while the first component will possess the largest distribution ratio h_1^0 and the largest velocity v_1^0.

Then let us show that the velocities $v_1^0, \ldots, v_j^0$ are only the initial velocities. Of these velocities, only one remains constant, undergoing no changes—the velocity of the j-th concentration point, the most sorbed component.

Let us take some i-th component for consideration and compare its distribution with the distribution of the j-th component. In an infinitesimal time interval dt, the front of the j-th component moves over a distance $v_j^0 dt$. Other components of the mixture, including the i-th component, should move together with the j-th component.

During the time dt, an amount of the i-th substance equal to $n_i^0 u\, dt$ enters the column. But could this amount of the component be accommodated in the zone of the j-th substance, the width of which is equal to $v_j^0 dt$? Let us determine the degree to which the i-th substance can be accommodated in the zone of the j-th substance. This amount will be equal to $(n_i^0 + N_i^0)\, v_j^0 dt$. Let us show that this amount of the i-th component will be smaller than the amount of the introduced i-th substance, i.e., $(n_i^0 + N_i^0)\, v_j^0 dt < n_i^0 u\, dt$.

According to formulas (IV.24), the latter inequality is identical with the inequality $v_j^0 < v_i^0$, i.e., from the condition $v_j^0 < v_i^0$, it unambiguously follows that the entire amount of the i-th substance introduced into the column cannot be accommodated in the zone of the j-th substance, with width $v_j^0 dt$. This excess amount of matter has only one way out—to penetrate beyond the zone of the j-th substance. Hence, all of the remaining, less sorbable components penetrate beyond the zone of the j-th substance in an infinitesimal time interval dt, i.e., theoretically, instantaneously.

Here lies the essence of the phenomenon of sorption displacement in the process of formation of a frontal chromatogram: the j-th, most sorbed substance displaces and pushes ahead all the less sorbed substances. But since in a mixture of substances that have penetrated beyond the zone of the j-th substance, the number of components of the mixture will be a unit smaller, i.e., equal to j−1, then naturally the equilibrium distribution of these substances should be instantaneously changed. Among the (j−1) components, situated beyond the zone of the j-th substance, new equilibrium concentrations are established, new distribution ratios, and new velocities. The least velocity among the (j−1) substances will be possessed by the (j−1)-th component, as the most sorbed among them. It should partially displace and push forward all the other (j−2) components. Then the picture will be analogous−sorption displacement of differently sorbed substances generates a series of infinitely narrow zones j, each successive one of which will contain one component less than the preceding, at the entrance to the column in an infinitesimal layer dx in an infinitesimal time dt, i.e., theoretically instantaneously. In front of all the components is formed the zone of the first, least sorbed substance. The concentrations of the components will differ in all these infinitely narrow zones. Thus, at the entrance to the column when x→0, a multitude of concentrations of the components arises simultaneously, in accord with the picture presented of the beginning of formation of a frontal chromatogram.

The theory of equilibrium frontal sorption dynamics of a mixture gives an idealized picture of the instantaneous formation of the frontal chromatogram. In a finite time of the process of sorption dynamics, this instantaneously arising frontal chromatogram will expand−chromatographic zones of finite width; boundaries will be formed−the fronts between the zones will move at the speeds characteristic of each boundary.

Let us give a more detailed description of the frontal chromatogram. The zonal structure of such a chromatogram and the compositions of the zones are schematically shown in Fig. 14.

At the beginning of the column, the most sorbed j-th substance forms a saturated zone with a straight break in the front. In this sorption zone, the j-th substance is exhausted (see Fig. 14). But, according to the sorption isotherm, in this zone the remaining components of the mixture will also exist in equilibrium with the sorbent. Thus, the highest j-th zone will be saturated with all the j-components:

$$N^0_{i,j} = f_{i,j}(n^0_{1,j},\ n^0_{2,j}, \ldots, n^0_{j,j}), \tag{IV.34}$$

where the first subscript denotes the number of the component, while the second denotes the number of the zone. In this expression, $n^0_{1,j}, n^0_{2,j}, \ldots, n^0_{j,j}$ are the equilibrium concentrations of the substances in the mobile phase in the j-th zone, equal to the concentrations in the initial mixture; $N^0_{1,j}, N^0_{2,j}, \ldots, N^0_{j,j}$ are the equilibrium concentrations in the sorbent for the j-th zone in accord with the isotherms (IV.34).

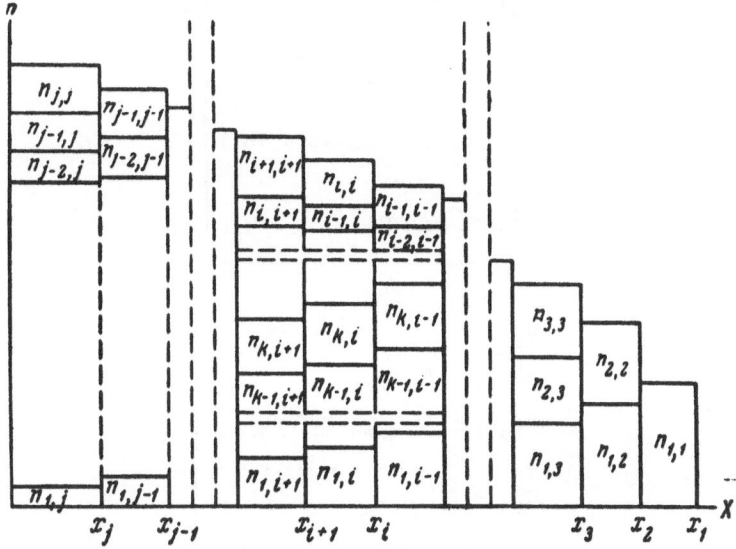

Fig. 14. Frontal chromatogram of a mixture of substances during their interdependent equilibrium sorption dynamics.

Since the sorption of the j-th component is exhausted in the j-th zone, (j–1) components go on past the boundary of this zone, the (j–1)-th component being the most sorbed component among them according to the series (IV.2). Its sorption will be exhausted in the following (j–1)-th zone.

In view of the fact that there is no j-th component in the (j–1)-th zone, and it thereby does not compete for sorption sites, the concentrations of the components contained in the (j–1)-th zone will be higher than in the j-th zone:

$$n_{i,\,i-1} > n^0_{i,\,j}, \ N_{i,\,i-1} > N^0_{i,\,j-1},$$

where $1 \le i \le j-1$. The distribution of the components between the mobile phase and the sorbent will obey the sorption isotherm:

$$N_{i,\,j-1} = f_{i,\,j-1}\ (n_{1,\,j-1}, n_{2,\,j-1}, \ldots, n_{j-1,\,j-1}). \qquad (\text{IV.35})$$

Since the sorption of the (j–1)-th component is exhausted in the (j–1)-th zone, (j–2) components will go beyond this zone, etc.

Hence, in each successive zone there will be one component fewer than in the preceding zone, while the equilibrium concentrations will be greater.

In general, for the i-th zone

$$N_{k,\,i} = f_{k,\,i}\ (n_{1,\,i}, \ n_{2,\,i}, \ \ldots, \ n_{i,\,i}), \ 1 \le k \le i. \qquad (\text{IV.36})$$

Finally, the first zone, the farthest forward, will contain only one component, the least sorbed—the first component in the sorption series (IV.2).

For it

$$N_{1,1} = f_{1,1}\ (n_{1,1}). \qquad (\text{IV.37})$$

The distribution of each component along the column is stepwise in character. A series of concentrations in the zones of the frontal chromatogram can be written for each component.

For the mobile phase

$$\left.\begin{aligned}
&n^0_{j,\,j} \\
&n^0_{j-1,\,j} < n_{j-1,\,j-1} \\
&n^0_{j-2,\,j} < n_{j-2,\,j-1} < n_{j-2,\,j-2} \\
&\quad \cdots \cdots \cdots \cdots \\
&n^0_{i,\,j} < n_{i,\,j-1} < \ldots < n_{i,\,i} \\
&\quad \cdots \cdots \cdots \cdots \\
&n^0_{1,\,j} < n_{1,\,j-1} < \ldots < n_{1,\,1}
\end{aligned}\right\} \qquad (\text{IV.38})$$

for the sorbent

$$\left.\begin{aligned}
&N^0_{j,\,j} \\
&N^0_{j-1,\,j} < N_{j-1,\,j-1} \\
&N^0_{j-2,\,j} < N_{j-2,\,j-1} < N_{j-2,\,j-2} \\
&\quad \cdots \cdots \cdots \cdots \\
&N^0_{i,\,j} < N_{i,\,j-1} < \ldots < N_{i,\,i} \\
&\quad \cdots \cdots \cdots \cdots \\
&N^0_{1,\,j} < N_{1,\,j-1} < \ldots < N_{1,\,1}
\end{aligned}\right\} \qquad (\text{IV.39})$$

From the analysis of the picture of formation of the frontal chromatogram cited, it follows that only the first zone—the zone of the least sorbed substance—will consist of one pure component. All the other zones will be mixed. Thus, the method of frontal chromatography does not give complete separation of a mixture of substances. Only part of the least sorbed component can be obtained in pure form. The method of frontal chromatography gives a partial separation of substances.

Thanks to sorption displacement, as can be seen from the series (IV.38), (IV.39), there is an increase in the concentrations of the components from zone to zone, which can be considered as the phenomenon of sorption enrichment.

A further problem consists of determining the concentrations of the substances in each zone, knowing the initial concentrations of the substances (IV.22) in the mobile phase and the sorption isotherm (II.6). This problem is solved on the basis of the law of conservation of matter.

Let us turn to Fig. 14. The rate of motion of the forward boundary of each zone is determined by the rate of motion of the front of the most strongly sorbed component contained in the given zone. The boundary between the j-th and (j−1)-th zones will move at the velocity of the front of the j-th component

$$v_j^0 = v_j = u \, \frac{n_{j,\,j}^0}{n_{j,\,j}^0 + N_{j,\,j}^0}, \tag{IV.40}$$

the boundary between the (j−1)-th and (j−2)-th zones will move at the velocity of the front of the (j−1)-th component in the (j−1)-th zone

$$v_{j-1} = u \, \frac{n_{j-1,\,j-1}}{n_{j-1,\,j-1} + N_{j-1,\,j-1}}, \tag{IV.41}$$

etc. In general, for the boundary between the i-th and (i−1)-th zones

$$v_i = u \, \frac{n_{i,\,i}}{n_{i,\,i} + N_{i,\,i}}, \tag{IV.42}$$

$$1 \leqslant i \leqslant j.$$

The width of the i-th zone will be

$$x_i - x_{i+1} = (v_i - v_{i+1})\,t = \left(\frac{n_{i,\,i}}{n_{i,\,i} + N_{i,\,i}} - \frac{n_{i+1,\,i+1}}{n_{i+1,\,i+1} + N_{i+1,\,i+1}} \right) ut. \tag{IV.43}$$

A relationship can be established between the concentrations of the components in two neighboring zones on the basis of the same law of conservation of matter.

Let us consider the balance of the k-th component between two cross sections, one of which, $L_1 L_2$ passes through the (i + 1)-th zone, while the second, $L_1' L_2'$ passes through the i-th zone (Fig. 15).

An amount of the k-th component equal to $n_{k,i+1}u\delta t$ passed through the cross section $L_1 L_2$ into the space between the indicated cross sections in the time δt. During this same time, an amount of the k-th component equal to $n_{k,i}u\delta t$ passed from this space through the cross section $L_1' L_2'$.

Since, as was shown above, $n_{k,i+1} < n_{k,i}$ [see the series (IV.38)], then $n_{k,i+1}u\delta t < n_{k,i}u\delta t$, and the amount of the k-th component decreased between the cross sections by the quantity

$$(n_{k,\,i} - n_{k,\,i+1})\,u\delta t. \tag{IV.44}$$

This decrease in the amount of the k-th component between the cross sections led to motion of the boundary between the zones at a velocity

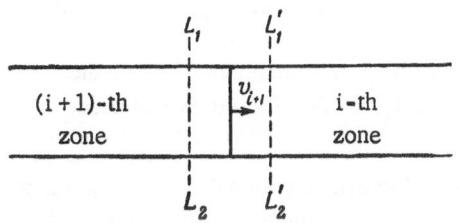

Fig. 15. For derivation of the recurrent formula (IV.49).

$$v_{i+1} = u \frac{n_{i+1,\,i+1}}{n_{i+1,\,i+1} + N_{i+1,\,i+1}} \qquad \text{(IV.45)}$$

over a distance $v_{i+1}\delta t$.

The decrease in the k-th component on this portion $v_{i+1}\delta t$ is equal to

$$[(n_{k,\,i} + N_{k,\,i}) - (n_{k,\,i+1} + N_{k,\,i+1})]\, v_{i+1}\delta t. \qquad \text{(IV.46)}$$

Hence, summarizing the balance of the k-th component, i.e., setting expressions (IV.44) and (IV.46) equal to each other, after reduction by δt, we obtain

$$(n_{k,\,i} - n_{k,\,i+1})\, u = [(n_{k,\,i} + N_{k,\,i}) - (n_{k,\,i+1} + N_{k,\,i+1})\, v_{i+1}. \qquad \text{(IV.47)}$$

Eliminating the quantity v_{i+1}/u from equations (IV.45) and (IV.47), we shall have

$$n_{k,\,i} - n_{k,\,i+1} = [(n_{k,\,i} + N_{k,i}) - (n_{k,\,i+1} + N_{k,\,i+1})] \frac{n_{i+1,\,i+1}}{n_{i+1,\,i+1} + N_{i+1,\,i+1}}. \qquad \text{(IV.48)}$$

After algebraic simplification of the last equation, we obtain the following recurrent system of equations:

$$n_{k,\,i} - n_{k,\,i+1} = \frac{n_{i+1,\,i+1}}{N_{i+1,\,i+1}} (N_{k,\,i} - N_{k,\,i+1}), \qquad \text{(IV.49)}$$

$$1 \leqslant k \leqslant i.$$

The following equations of the sorption isotherms should be substituted into formula (IV.49):

$$N_{k,\,i} = f_{k,\,i}\, (n_{1,\,i}, n_{2,\,i}, \ldots, n_{i,\,i}); \qquad \text{(IV.50)}$$

$$N_{k,\,i+1} = f_{k,\,i+1}\, (n_{1,\,i}, \ldots, n_{i+1,\,i+1}); \qquad \text{(IV.51)}$$

$$N_{i+1,\,i+1} = f_{i+1,\,i+1}\, (n_{1,\,i+1}, \ldots, n_{i+1,\,i+1}). \qquad \text{(IV.52)}$$

The system of equations (IV.49) will consist of i equations with i unknown concentrations $n_{1,i}, n_{2,i}, \ldots, n_{1,i}$ for the i-th zone. This system is solvable in principle, if the concentrations of the components in the preceding $(i+1)$-th zone are known. Here lies the recurrent character of the system (IV.49). Since the concentrations of the components of the mixture are known for the highest, j-th zone (initial concentrations), then the concentrations for all the zones of the primary chromatogram, the rates of motion of the zones (IV.42), and their widths (IV.43) can gradually be calculated.

In order to represent once again the initial (at $t = 0$) state of the equilibrium frontal chromatogram of a mixture of substances with all the instantaneously established concentrations of the substances and rates of their motion, we must mentally compress the picture of the distribution of the substances depicted in Fig. 14 to an infinitesimally narrow width. This gives us an idea of the multitude of concentrations at the entrance to the column at the initial moment. Consequently, if we use a more rigorous approach, then the boundary conditions must include the entire series of concentrations instantaneously generated at the entrance, as written in the series (IV.38) and (IV.39).

As is well-known, the method of frontal chromatography can be used for analytical purposes. The problem of quantitative analysis of a mixture of substances is opposite to the problem that we were considering thus far—the quantitative composition of the initial mixture of substances must be determined according to the frontal chromatogram. Such a problem is solved by the method developed by A. Tiselius and S. Claesson, which has received the name of frontal analysis [228, 77].

The essence of frontal analysis consists of the following. The so-called effluent frontal chromatogram is obtained—the concentrations of the substances are measured at the exit from the column. Quantitative analysis of the effluent chromatograms using suitable methods of analysis is an incomparably easier problem in the experimental respect than the analysis of substances in the column.

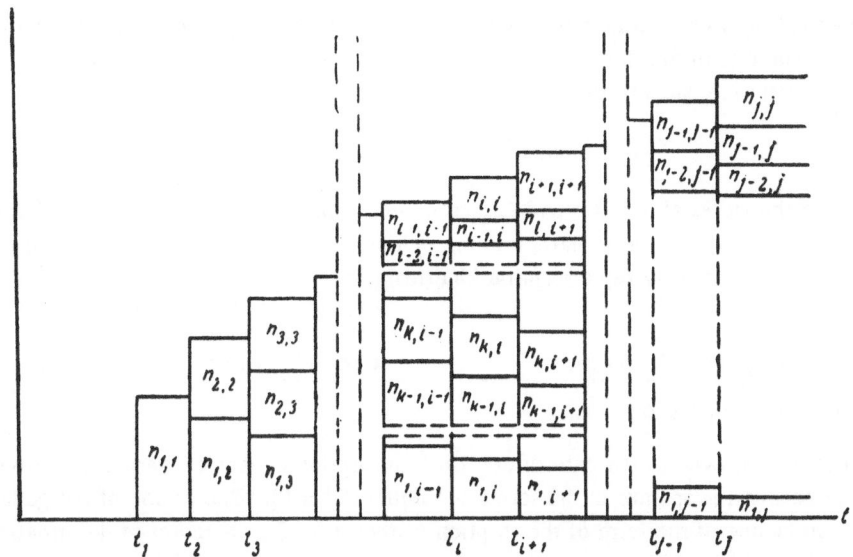

Fig. 16. Effluent frontal chromatogram of a mixture of substances. Frontal analysis of a mixture of substances.

Moreover, it is possible to show theoretically on the basis of an analysis of equations (IV.49) that it is sufficient to determine the summary concentrations of the substances in the zones of the effluent frontal chromatogram. A suitable method of quantitative analysis can be selected, with the aid of which the summary concentration of the substances in the zones of the substances successively emerging from the column will be measured. The effluent curve thereby obtained will also represent a stepwise distribution curve (Fig. 16).

The first component emerges first from the column (first zone), then two components (second zone), etc. At the end of the process, the initial mixture of substances with the initial concentrations will emerge (j-th zone). The first step will consist of the pure first component. More briefly, the following quantities will be determined experimentally:

$$\left.\begin{aligned}
\Sigma_1 &= n_{1,1}; \\
\Sigma_2 &= n_{1,2} + n_{2,2}; \\
&\cdot \cdot \cdot \cdot \cdot \cdot \cdot \cdot \cdot \cdot \cdot \\
\Sigma_i &= n_{1,i} + n_{2,i} + \ldots + n_{i,i}; \\
&\cdot \cdot \cdot \cdot \cdot \cdot \cdot \cdot \cdot \cdot \cdot \\
\Sigma_j &= n_{1,j}^0 + n_{2,j}^0 + \ldots + n_{j,j}^0.
\end{aligned}\right\} \qquad \text{(IV.53)}$$

The same recurrent system of equations (IV.49) can be used to calculate the concentrations in each of the zones, considering the concentrations in the i-th zone as the known quantities in it, and the concentrations in the (i + 1)-th zone as the unknowns.

But we have i equations of the type of (IV.49), while the number of unknown quantities is (i + 1). We need still another equation for the system to be solvable.

Let us take the following as the (i + 1)-th equation:

$$n_{i+1,\, i+1} = \Sigma_{i+1} - (n_{1,\, i+1} + n_{2,\, i+1} + \ldots + n_{i,\, i+1}), \qquad \text{(IV.54)}$$

i.e., the concentration $n_{i+1,\, i+1}$ is determined according to the difference between the summary concentration of all the components in the (i + 1)-th zone Σ_{i+1} and the summary concentration of the i components in the same zone.

The concentration $n_{1,1}$ is determined analytically—this is the first step in the effluent curve. But then begins the calculation according to the system of equations (IV.49), (IV.54). The concentrations $n_{1,1}$ and $n_{1,2}$ in the second zone (second step) are calculated; after determining them, we pass on to a calculation of the concentrations $n_{1,3}$, $n_{2,3}$ and $n_{3,3}$ in the third zone (third step), etc. Thus, we gradually reach the very last j-th zone (j-th step), in which the concentrations of the components are equal to the initial concentration sought:

$$n_{1,j}^0 = n_1^0, \; n_{2,j}^0 = n_2^0, \; \ldots, \; n_{j,j}^0 = n_j^0.$$

As was shown in the works of Tiselius and Claesson and in many others, in a number of cases the boundaries between the zones obtained are very sharp, and the dynamics of sorption of substances are realized under conditions close to equilibrium (very rapid sorption kinetics).

3. Equilibrium Sorption Dynamics of a Mixture of Substances under the Action of Longitudinal Effects ($D^* \neq 0$)

Under real conditions, even in the case of equilibrium sorption dynamics, various longitudinal effects of blurring of the zone boundaries operate in the sorption column ($D^* \neq 0$). The action of the general principle of sorption dynamics—influence of the form of the sorption isotherm on the character of deformation of the fronts—is thereby manifested.

The case when all the isotherms are convex is of the general greatest practical interest. In this case, the factor of convexity of the sorption isotherms of a mixture of substances is a factor of compression and stabilization of the fronts of sorption dynamics. The frontal chromatogram of a mixture of substances thereby will have the same structure as when the action of blurring longitudinal effects is absent ($D^* = 0$)—the number of zones with the same composition. The only difference will lie in the fact that when $D^* \neq 0$, the boundaries between zones will be blurred. In the case of convex sorption isotherms of the substances, stationary blurred fronts, moving at a constant rate, are formed at some asymptotic stage of the sorption dynamics.

The theory of equilibrium frontal chromatography, outlined in the preceding section, is entirely applicable to the calculation of the concentrations of the components in the zones of saturation. If the blurring of the fronts is very negligible (the widths of the blurred fronts is considerably smaller than the zones of saturation), then the same theory can be used for the approximate evaluation of the widths of the chromatographic zones. Using the Zel'dovich-Todes method, the distribution of substances in the stationary fronts when $D^* \neq 0$ can in principle be calculated.

Figure 17 schematically shows a portion of a frontal chromatogram in the case of steady-state blurring of the fronts. Let us assume that the equilibrium concentrations of the components in all the zones of the chromatogram were calculated using the recurrent system of equations (IV.49). Let us take for consideration the transition region—the region of the steady-state fronts—from the i-th to the (i−1)-th zone. Under the conditions of a steady-state system, the fronts of all i components should move at the same constant velocity v_i—the rate of motion of the front of the component most sorbed in the given zone, which it "ties" to all the other, less sorbed components.

Equilibrium sorption dynamics at $D^* \neq 0$ should be described by the equations of balance:

$$\frac{\partial n_k}{\partial t} + u \frac{\partial n_k}{\partial x} + \frac{\partial N_k}{\partial t} = D_k^* \frac{\partial^2 n_k}{\partial t^2}, \qquad \text{(IV.54')}$$

$$1 \leqslant k \leqslant i,$$

in which n_k and N_k are the equilibrium concentrations of the components in the fronts, while the equations of the sorption isotherms:

$$N_k = f_k (n_1, n_2, \ldots, n_i), \; 1 \leqslant k \leqslant i. \qquad \text{(IV.55)}$$

Fig. 17. Blurring of the boundaries between zones of the frontal chromatogram in the case of interdependent sorption in a real chromatographic column.

Let us introduce a new variable

$$z_i = x - v_i t \qquad \text{(IV.56)}$$

and transform equation (IV.54') to a new system of coordinates z_i:

$$-v_i \frac{dn_k}{dz_i} + u \frac{dn_k}{dz_i} - v_i \frac{dn_k}{dz_i} = D_k^* \frac{d^2 n_k}{dz_i^2} . \qquad \text{(IV.57)}$$

For the transition region selected from the i-th to the (i−1)-th zone, we shall have the following boundary conditions:

$$z_i \rightarrow + \infty, \ n_i = 0, \ N_i = 0, \ \frac{dn_i}{dz} = 0; \qquad \text{(IV.58)}$$

$$n_k = n_{k,\,i-1}, \ N_k = N_{k,\,i-1}, \ \frac{dn_k}{dz} = 0, \qquad \text{(IV.59)}$$
$$1 \leqslant k \leqslant i - 1;$$

$$z_i \rightarrow - \infty, \ n_i = n_{i,\,i}, \ N_i = N_{i,\,i}, \ \frac{dn_i}{dz} = 0; \qquad \text{(IV.60)}$$

$$n_k = n_{k,\,i}, \ \ N_k = N_{k,\,i}, \ \frac{dn_k}{dz} = 0, \qquad \text{(IV.61)}$$
$$1 \leqslant k \leqslant i.$$

Here it was necessary to substitute the separate boundary condition for the i-th component, in which the concentrations are equal to zero when $z \rightarrow + \infty$, for the remaining components present in both neighboring zones as well.

Correspondingly, after integration of equation (IV.57), we obtain the following equations:

$$D_i^* \frac{dn_i}{dz_i} = (u - v_i) \, n_i - v_i N_i + C_i; \qquad \text{(IV.62)}$$

$$D_k^* \frac{dn_k}{dz_i} = (u - v_i) \, n_k - v_i N_k + C_k, \qquad \text{(IV.63)}$$
$$1 \leqslant k \leqslant i - 1,$$

in which the integration constants C_i and C_k will differ, depending on the boundary conditions. The intergration constant C_i is determined from the condition (IV.58). We obtain $C_i = 0$. Hence

$$D_i^* \frac{dn_i}{dz_i} = (u - v_i) \, n_i - v_i N_i. \qquad \text{(IV.64)}$$

Substituting the boundary condition (IV.60) into the last equation, we obtain

$$(u - v_i) \, n_{i,\,i} - v_i N_{i,\,i} = 0, \qquad \text{(IV.65)}$$

from which the rate of motion of the steady-state fronts can be determined:

$$v_i = u \frac{n_{i,\,i}}{n_{i,\,i} + N_{i,\,i}} , \qquad \text{(IV.42)}$$

which coincides with the already known formula (IV.42).

Let us determine the constant C_k. For this, let us substitute the boundary conditions (IV.59) and (IV.61) into equation (IV.63)

$$(u - v_i)\, n_{k,\,i-1} - v_i N_{k,\,i-1} + C_k = 0, \qquad \text{(IV.66)}$$

$$(u - v_i)\, n_{k,\,i} - v_i N_{k,\,i} + C_k = 0, \qquad \text{(IV.67)}$$

from which

$$C_k = -\left[(u - v_i)\, n_{k,\,i-1} - v_i N_{k,\,i-1}\right] = -\left[(u - v_i)\, n_{k,\,i} - v_i N_{k,\,i}\right]. \qquad \text{(IV.68)}$$

From the last equation, we can also determine the ratio of motion of the steady-state fronts:

$$v_i = u\,\frac{n_{k,\,i-1} - n_{k,\,i}}{(n_{k,\,i-1} - n_{k,\,i}) + (N_{k,\,i-1} - N_{k,\,i})}. \qquad \text{(IV.69)}$$

Setting the right-hand portions of formulas (IV.42) and (IV.69) equal, we obtain the function

$$\frac{n_{k,\,i-1} - n_{k,\,i}}{N_{k,\,i-1} - N_{k,\,i}} = \frac{n_{i,\,i}}{N_{i,\,i}}, \qquad \text{(IV.70)}$$

which coincides with the function (IV.49), already known earlier. Now let us substitute into equation (IV.63) one of the expressions for the integration constant C_k according to (IV.68). Then we obtain

$$D_k^* \frac{dn_k}{dz_i} = (u - v_i)\,(n_k - n_{k,\,i}) - v_i\,(N_k - N_{k,\,i}). \qquad \text{(IV.71)}$$

The system of equations (IV.64), (IV.55), and (IV.71), consisting of i equations with i unknown concentrations n_k, describes the sorption dynamics of the i components in the steady-state fronts between the i-th and (i−1)-th zones, and can be used in principle to calculate the distribution of the components in these fronts.

Of course, this is only the general formulation of the problem. The solution of such a system in general form is impossible. But particular solutions can be obtained in individual cases.

Since the forward front of the first zone will contain only one, the first component, the contour of this front can be calculated on the basis of the equations of sorption dynamics of one substance. Such a calculation can be utilized in practice, for example, for evaluating the time of beginning of emergence of the first component from the column of the sorbent, which is very important when sorbent columns are used as sorption filters for trapping polycomponent mixtures of substances.

4. Nonequilibrium Sorption Dynamics of a Mixture of Substances

The calculation of the distribution of a mixture of substances to be sorbed under nonequilibrium conditions in the case of independent sorption of substances (linear isotherm) is performed for each individual component according to the theory of nonequilibrium sorption dynamics of one substance.

In the case of interdependent sorption, the character of the blurring of the fronts under the action of the kinetic factor will depend on the type of the sorption isotherm (convex or concave isotherm). Only in the case of a convex sorption isotherm is there stabilization of the fronts, which will move at a constant rate, at the asymptotic stage of sorption dynamics.

The equilibrium concentrations of the substances in the regions of saturation can be calculated according to the recurrent equations (IV.49). The calculation of the contours of the fronts can also be performed in principle by the Zel'dovich-Todes method.

Let us turn again to Fig. 17, which schematically shows the character of the distribution of substances in a frontal chromatogram under conditions of steady-state blurring of the fronts. Let us consider the transition blurred region between the i-th and (i−1)-th zones.

Let us write a system of differential equations of nonequilibrium sorption dynamics for i components of the i-th zone at $D^* = 0$:

$$\frac{\partial n_k}{\partial t} + u\,\frac{\partial n_k}{\partial x} + \frac{\partial N_k}{\partial t} = 0, \qquad (IV.72)$$

$$\frac{\partial N_k}{\partial t} = \psi_k\,(n_1,\ n_2,\ \dots,n_i,\ N_1,\ \dots,N_i),$$
$$1 \leqslant k \leqslant i. \qquad (IV.73)$$

At the asymptotic stage of sorption dynamics, the fronts of all the components will move at a constant velocity v_i. Let us transform equations (IV.72) and (IV.73) to a new system of coordinates through the substitution of (IV.56):

$$-\,v_i\frac{dn_k}{dz_i} + u\,\frac{dn_k}{dz_i} - v_i\frac{dN_k}{dz_i} = 0; \qquad (IV.74)$$

$$-\,v_i\frac{dN_k}{dz_i} = \psi_k\,(n_1,\ \dots,\ n_i,\ N_1,\ \dots,\ N_i). \qquad (IV.75)$$

The boundary conditions will take the form:

$$z_i \to +\infty,\ \ n_i = 0,\ \ \ N_i = 0; \qquad (IV.76)$$

$$n_k = n_{k,\,i-1},\ \ N_k = N_{k,\,i-1}, \qquad (IV.77)$$
$$1 \leqslant k \leqslant i - 1;$$

$$z_i \to -\infty,\ \ n_i = n_{i,\,i},\ \ N_i = N_{i,\,i}; \qquad (IV.78)$$

$$n_k = n_{k,\,i},\ \ N_k = N_{k,\,i}, \qquad (IV.79)$$
$$1 \leqslant k \leqslant i.$$

For the i-th component, after integration of equation (IV.74), we obtain the following solution:

$$(u - v_i)\,n_i - v_iN_i = C_i, \qquad (IV.80)$$

and for the remaining components:

$$(u - v_i)\,n_k - v_iN_k = C_k,$$
$$1 \leqslant k \leqslant i - 1. \qquad (IV.81)$$

The integration constants C_i and C_k will depend on the boundary conditions (IV.76), (IV.79). Substituting the boundary conditions (IV.76) into equation (IV.80), we obtain $C_i = 0$. Hence, we have

$$(u - v_i)\,n_i - v_iN_i = 0. \qquad (IV.82)$$

Substitution of the boundary condition (IV.78) into equation (IV.82) gives

$$(u - v_i)\,n_{i,\,i} - v_iN_{i,\,i} = 0. \qquad (IV.83)$$

From equations (IV.82) and (IV.83), we obtain the formula for the rate of motion of the steady-state fronts:

$$v_i = u\,\frac{n_i}{n_i + N_i} = u\,\frac{n_{i,\,i}}{n_{i,\,i} + N_{i,\,i}}, \qquad (IV.84)$$

as well as the linear relationship between the nonequilibrium concentrations of the i-th component in the front:

$$\frac{n_i}{N_i} = \frac{n_{i,\,i}}{N_{i,\,i}}. \qquad (IV.85)$$

Thus, the linear Zel'dovich function for the i-th component is also fulfilled in the case of sorption dynamics of a mixture of substances. However, the linear relationship between the nonequilibrium concentrations in the front, in the form of function (IV.85), is fulfilled only for the i-th component, which is absent in the (i–1)-th zone.

For the remaining (i–1) components, a linear relationship will also be observed among the nonequilibrium concentrations in the fronts. However, the form of such a dependence will be somewhat different.

Substitution of the boundary conditions (IV.77) and (IV.79) into equation (IV.81) gives

$$\left.\begin{array}{l} (u - v_i)\, n_{k,i-1} - v_i N_{k,i-1} = C_k; \\ (u - v_i)\, n_{k,i} - v_i N_{k,i} = C_k. \end{array}\right\} \tag{IV.86}$$

From equations (IV.81) and (IV.86), after elimination of the constant C_k, we obtain the following equations:

$$(u - v_i)\,(n_k - n_{k,i}) - v_i\,(N_k - N_{k,i}) = 0; \tag{IV.87}$$

$$(u - v_i)\,(n_k - n_{k,i-1}) - v_i\,(N_k - N_{k,i-1}) = 0; \tag{IV.88}$$

$$(u - v_i)\,(n_{k,i-1} - n_{k,i}) - v_i\,(N_{k,i-1} - N_{k,i}) = 0. \tag{IV.89}$$

Expressions for the rate of motion of the steady-state fronts can also be obtained from these expressions:

$$v_i = u\,\frac{n_k - n_{k,i}}{(n_k - n_{k,i}) + (N_k - N_{k,i})} = u\,\frac{n_k - n_{k,i-1}}{(n_k - n_{k,i-1}) + (N_k - N_{k,i-1})} = u\,\frac{n_{k,i-1} - n_{k,i}}{(n_{k,i-1} - n_{k,i}) + (N_{k,i-1} - N_{k,i})}$$

$$\tag{IV.90}$$

Setting the last expression in (IV.90) equal to the last expression in (IV.84), we obtain the function (IV.70), coinciding with the function (IV.49), which we already know. On the other hand, from equations (IV.87)–(IV.89) we obtain the following functions:

$$\frac{n_k - n_{k,i}}{N_k - N_{k,i}} = \frac{n_k - n_{k,i-1}}{N_k - N_{k,i-1}} = \frac{n_{k,i-1} - n_{k,i}}{N_{k,i-1} - N_{k,i}}. \tag{IV.91}$$

The functions (IV.91) obtained give us the relationship among the nonequilibrium concentrations in the steady-state fronts sought:

$$N_k = \frac{N_{k,i} n_{k,i-1} - n_{k,i} N_{k,i-1}}{n_{k,i-1} - n_{k,i}} + \frac{N_{k,i-1} - N_{k,i}}{n_{k,i-1} - n_{k,i}}\, n_k. \tag{IV.92}$$

As can be seen, this function is linear, and in a more general form represents the Zel'dovich linearity function. Actually, at $k = i$, equation (IV.92) is transformed to (IV.85), since the i-th component is absent in the (i–1)-th zone:

$$n_{i,i-1} = 0, \qquad N_{i,i-1} = 0.$$

Their distribution of the components in the fronts of the chromatogram $f(n_k, x, t)$ and $f(N_k, x, t)$ can in principle be calculated on the basis of the solution of the system of the kinetic equations (IV.75). The value of the velocity according to formula (IV.84) should be substituted into this system, and in place of N_i and N_k — their values from the functions (IV.85) and (IV.92).

The solution of such a scheme of differential equations in general form is of course impossible. However, particular solutions can be obtained with suitable selection of the sorption kinetic equations.

As yet, only one case is known of an approximate solution of the problem of nonequilibrium sorption dynamics of two substances [6, 8]. Since the first zone of the frontal chromatogram contains only one component, the contour of the forward front for this zone can be calculated according to the theory of the steady-state front of nonequilibrium sorption dynamics of one substance.

CHAPTER V

THEORY OF ELUTION CHROMATOGRAPHY

1. Elution Sorption Dynamics of One Substance

__Equilibrium Sorption Dynamics.__ Let the process of elution of the primary (frontal) zone of one substance be set by the following initial and boundary conditions:

$$t = 0, \ 0 \leqslant x \leqslant x_0, \ n = n_0, \ N = N_0; \ \Big\} \text{(primary zone)} \tag{V.1}$$
$$x > x_0, \ n = 0, \quad N = 0;$$

$$t > 0, \qquad x = 0, \ n = 0, \ N = 0; \ \Big\} \text{(elution)} \tag{V.2}$$
$$x = \infty, \ n = 0, \ N = 0.$$

The problem consists of finding the course of the motion and distribution of the substance to be sorbed during the process of elution of the column with a pure solvent (mobile carrier), after the primary zone has been obtained. Here let us assume that the same solvent (or mobile carrier gas) from which the primary zone was formed was used for the elution.

The solution of the problem posed can be given on the basis of the Wilson law. If no perturbations occur on the boundaries of the primary zone, and the concentrations $n = n_0$ and $N = N_0$ are kept constant, then in filtration through the column of pure solvent, the primary zone will be displaced at a constant rate $v = uh/(1+h)$, without changing its original width x_0 [241]. This is explained by the fact that since there is only one concentration in the zone, n_0, and the concentration N_0 related to it by the sorption isotherm, then, according to the Wilson law, only one rate of displacement of this concentration should exist. Hence, the points n_0 and N_0 at the front and rear boundaries of the zone will move at the same rate $v = uh/(1+h)$—the width of the zone will be preserved.

When perturbations exist at the boundaries of the zone, disturbing the constancy of the concentrations, in the case of equilibrium sorption dynamics, according to the Wicke law, there will be a deformation of the forward and rear fronts, depending on the type of the isotherm [188, 189]. According to the Wicke law (III.8), in the case of a convex isotherm, the forward front should be compressed, while the rear front should be stretched out. Thus, in this case any perturbation at the forward front should soon be liquidated as a result of the phenomenon of sharpening of the forward front. In the case of a convex isotherm, the forward front should be sharper, and if there are no systematic perturbations, the front will have a "straight" break (Fig. 18). The rate of its motion should be constant: $v = uh/(1+h)$.

As follows from the Wicke law (III.8), random perturbations at the rear front of the zone of the sorbed substance cannot be eliminated. On the contrary, a blurring of the rear front, having once arisen, will progress (formation of a "tail," see Fig. 18). Since each point of the rear front moves at its own characteristic constant velocity according to the formula (III.8), the blurring of the rear front will occur in proportion to the time t. Using formula (III.8), we can calculate the distribution of the substance to be sorbed in the rear front at various stages of elution.

At some stage of elution, as a result of the blurring of the rear front, the region of constant concentration (the "plateau" of the zone) disappears, and a maximum appears in the distribution curve.

For the forward front of the zone, from the moment of disappearance of the "plateau," the previous boundary conditions (behind the front) are disturbed; behind the front we shall no longer have $n_0 = $ const. Actually, the point of the maximum turns into the extreme upper concentration point of the rear front of the zone. This point

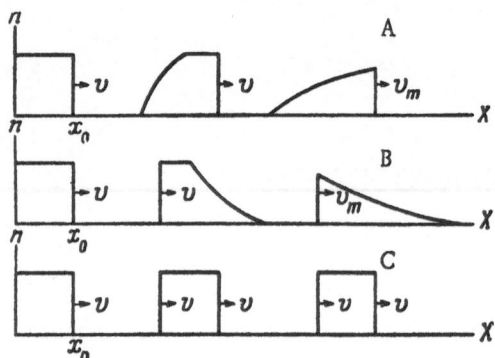

Fig. 18. Elution of the zone of one substance in the case of equilibrium sorption dynamics. A) Convex isotherm; B) concave isotherm; C) linear isotherm.

will now move according to Wicke's law (III.8). This means that at the stage after disappearance of the "plateau" of the zone, the rate of motion of the forward front v_m will be determined by the rate of motion of the concentration point of the maximum according to Wicke's law. The height of the maximum will gradually decrease. In the case of prolonged elution, the substance to be sorbed is "smeared out" along the column.

In the case of a convex isotherm, the picture of deformation of the latter and of the rear front will be opposite to the picture just considered—the forward front will be blurred (formation of a "tongue," see Fig. 18), while the rear front will be contracted being converted to a front with a "straight" break.

Let us note that from the practical standpoint, in the regeneration (purification) of the sorption column, the case of elution sorption dynamics with a concave isotherm possesses a valuable advantage over the case of elution dynamics with a convex isotherm. In the case of a concave isotherm, optimum conditions are created for conducting a rapid and effective regeneration of the sorption column.

In the case of a linear isotherm, a perturbation once arisen at the boundaries will be preserved. But if there are no perturbations, then the primary zone will move at a constant rate $v = uh/(1+h)$ without deformation (see Fig. 18).

In a real column of sorbent, even under equilibrium conditions of sorption, there will be constant perturbations, creating blurring, both on the forward and on the rear fronts ($D^* \neq 0$).

Under such conditions, in the case of a convex sorption isotherm, the forward front (the rear front in the case of a concave isotherm) will be blurred. However, in view of the convexity of the sorption isotherm, on the strength of equation (III.8), the front should be stabilized at some stage of elution (Fig. 19).

The theory of frontal sorption dynamics of one substance—formulas (III.58)-(III.61)—can be used to calculate the contour of the forward steady-state front. Since the rear front will be progressively blurred, at some stage of elution, the region of the zone where $n = n_0$ ("plateau") disappears. From this moment on, the boundary conditions of the steady-state character of the forward front are disturbed (there is no region $n = n_0$ behind); the forward front will also begin to be deformed, the height of the maximum will decrease, and as a result of blurring of the rear front, the substance will gradually be smeared out along the column.

In the case of a concave sorption isotherm, the rear front will be stabilized (see Fig. 19), and the forward front will be progressively blurred.

Let us consider the elution sorption dynamics in the case of a linear sorption isotherm. Let the primary zone be set in the form of a narrow band ($x_0 \to 0$). The blurring of such a zone during elution when $D^* \neq 0$ can be considered by analogy with the process of diffusion spreading of a narrow band of matter [11, 17, 163, 164]. The equilibrium sorption dynamics in the case of $D^* \neq 0$ and a linear isotherm is described by the equation (III.76) already considered. The solution of this equation for the asymptotic

Fig. 19. Elution of the zone of one substance under the action of longitudinal or kinetic factors of blurring. A) Convex isotherm; B) concave isotherm; C) linear isotherm.

step of the process of elution of the initial narrow band can be approximately described by the following equation:

$$\frac{n}{n_0} = \frac{N}{N_0} = \frac{1}{\sqrt{4\pi Ht}}\, e^{-\frac{(x-vt)^2}{4Ht}}. \tag{V.3}$$

This formula is analogous to the Gaussian formula of error distribution known in mathematical statistics. We obtain the maximum value of the relative concentration $\varphi = n/n_0$ when the exponent is equal to zero: $x_{max} - vt = 0$, from which the rate of motion of the maximum

$$v_{max} = \frac{x_{max}}{t} = v = u\,\frac{h}{1+h}. \tag{V.4}$$

The value of the maximum concentration $\varphi_{max} = n_{max}/n_0 = 1/\sqrt{4\pi Ht}$, or

$$n_{max} = \frac{n_0}{\sqrt{4\pi Ht}}, \tag{V.5}$$

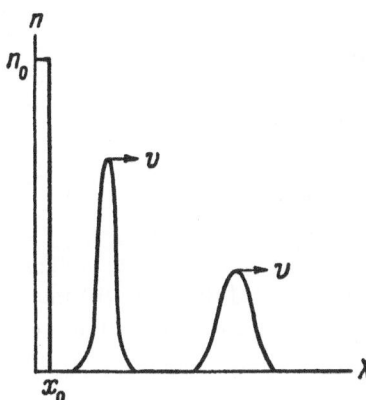

Fig. 20. Elution of a narrow zone of one substance under the action of longitudinal or kinetic factors of blurring in the case of a linear isotherm.

i.e., the maximum concentration decreases in inverse proportion to $\sqrt{t}$.

The motion and deformation of the zone during elution can be schematically depicted as shown in Fig. 20.

Let us conditionally determine the width of the zone at the level $n/n_{max} = e^{-1}$. At this concentration level, we can obtain the positions of the two concentration points from the equation

$$\frac{(x-vt)^2}{4Ht} = 1. \tag{V.6}$$

The coordinates sought, according to equation (V.6), will be:

$$x_1 = vt + 2\sqrt{Ht}; \tag{V.7}$$

$$x_2 = vt - 2\sqrt{Ht}. \tag{V.8}$$

The width of the zone, according to (V.7) and (V.8), will be

$$\delta_x = x_1 - x_2 = 4\sqrt{Ht}, \tag{V.9}$$

i.e., there will be expansion of the zone, proportional to $\sqrt{t}$.

In the case when a broad initial band of the substance to be sorbed is set, the process of deformation of the band during elution will consist of two stages. At the first stage, there is a blurring of the edges of the zone, which can be described within the framework of the theory of frontal sorption dynamics with a linear isotherm. The forward front is deformed according to the following asymptotic equation:

$$\frac{n}{n_0} = \frac{N}{N_0} = 0.5\left\{1 - \operatorname{erf}\left[\frac{(x-x_0)-vt}{2\sqrt{Ht}}\right]\right\}, \tag{V.10}$$

and the rear front according to:

$$\frac{n}{n_0} = \frac{N}{N_0} = 0.5\left[1 + \operatorname{erf}\left(\frac{x-vt}{2\sqrt{Ht}}\right)\right]. \tag{V.11}$$

A zone in the form of a band with blurred edges (Fig. 21) will move during elution at a rate $v = uh/(1+h)$, as follows from (V.10), (V.11).

The second stage begins after the region of the "plateau" where $n = n_0$ disappears in the zone. For this stage, the distribution of the substance in the zone can be approximately described by formula (V.3). Since in

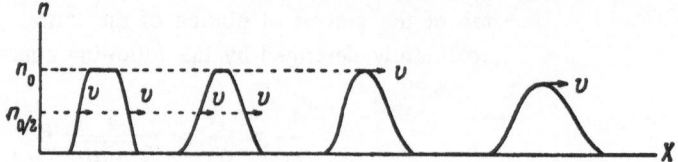

Fig. 21. Elution of a broad zone of one substance under the
action of longitudinal or kinetic factors of blurring in the case
of a linear isotherm.

the case of a linear isotherm, the partition coefficient h does not change, the maximum of the zone will move
as before at a rate $v_{max} = uh/(1+h)$. At all stages of the elution dynamics, expansion of the zone will occur in
proportion to $\sqrt{t}$.

Nonequilibrium Sorption Dynamics. Under the conditions of nonequilibrium sorption dynamics, the finite
rate of sorption leads to blurring of the boundaries of the zone. However, here too the deciding influence on
the process of elution sorption dynamics is exerted by the sorption isotherm. In the case of a convex isotherm,
the forward front should be sharpened, while the rear front should be stretched out.

The contour of the forward front can be calculated on the basis of the theoretical formulas for the steady-
state front [see formulas (III.96), (III.97), (III.98), (III.99)]. The contour of the rear front can be approximately
calculated according to the formula of equilibrium dynamics (III.8).

In the case of a concave isotherm, the picture of deformation of the zone will be the opposite. In the
case of nonequilibrium sorption dynamics, the course of deformation of the zones will be externally analogous
to the course of deformation in the case of equilibrium dynamics when $D^* \neq 0$. The same can be said with respect to the more complex case when both factors of blurring operate—longitudinal effects and sorption kinetics.

In the case of a linear sorption isotherm in the process of elution sorption dynamics, there is a symmetri-
cal, progressive blurring of the forward and rear fronts of the substance to be sorbed. This process can be considered within the framework of the semiphenomenological theories of hysteresis.

At the beginning let us assume that $D^* = 0$, and the blurring of the zone is due only to the sorption kinetics. Let the primary zone be set at the beginning of the column in the form of an infinitesimally narrow
band $(x_0 \to 0)$. Then, using the differential equations (III.102) or (III.103), for the elution of an infinitesimally
narrow band, we can obtain the following approximate asymptotic solutions by analogy with the solution of the
problems of diffusion [11, 154, 163, 164]:

$$\frac{N}{N_0} \approx \frac{n}{n_0} = \frac{1}{\sqrt{\frac{4\pi\varkappa}{1+h}vt}} e^{-\frac{(x-vt)^2}{\frac{4\varkappa}{1+h}vt}} ; \tag{V.12}$$

$$\frac{N}{N_0} \approx \frac{n}{n_0} = \frac{1}{\sqrt{\frac{4\pi\tau}{1+h}v^2t}} e^{-\frac{(x-vt)^2}{\frac{4\tau}{1+h}v^2t}} ; \tag{V.13}$$

or, if, according to (III.107), we introduce the quasidiffusion constant D_q^*, then

$$\frac{N}{N_0} \approx \frac{n}{n_0} = \frac{1}{\sqrt{4\pi D_q^* \frac{h}{1+h}t}} e^{-\frac{(x-vt)^2}{4D_q^* \frac{h}{1+h}t}} . \tag{V.14}$$

Equations (V.12)-(V.14) give practically equivalent results, independent of the true nature of the dynamic constants $\varkappa$, τ, and D_q^* only in the region of concentrations 10-90% of n_{max} [163, 164]. In the region of low concentrations, both in the forward and in the rear fronts, the nature of the mechanism of blurring begins to emerge (external or internal diffusion [163, 164]).

In the presence of longitudinal effects ($D_l^* \neq 0$), the generalized coefficient of quasidiffusion $D^* = D_q^* + D_l^*$ can be introduced. The nature of these longitudinal effects can be manifested appreciably in the region of low concentrations $n \ll n_{max}$. The case of blurring of the broad primary zone can be considered analogously to that for equilibrium sorption dynamics [see formulas (V.10), (V.11), and Fig. 21].

We should keep in mind here that under nonequilibrium conditions, there is no linear relationship among the nonequilibrium concentrations. However, for an asymptotic distribution, when the maximum of the zone travels a sufficiently long distance along the column, taking into consideration the smallness of the parameters $\varkappa$ and τ, we can approximately consider that $n/n_0 \approx N/N_0$. Moreover, it can be shown that the maxima N_{max} and n_{max} will be displaced with respect to one another by a quantity $\varkappa$ or τ [163, 164].

According to (V.12)-(V.14), in the case of nonequilibrium sorption dynamics, in the case of a linearity of the sorption isotherm, the width of the zone during elution of the column increases in proportion to $\sqrt{t}$. The concentration maxima n_{max} and N_{max} decrease continuously with time. Only the constancy of their rate of motion $v = uh/(1+h)$ is preserved.

2. Elution Sorption Dynamics of a Mixture of Substances

In the case of independent sorption of the components, when the sorption isotherms of the individual components are linear, in the case of equilibrium elution sorption dynamics, each component will move independent of others, with its own constant velocity

$$v_i = u\,\frac{h_i}{1+h_i}\,, \qquad (V.15)$$

where h_i is the partition coefficient of the i-th component.

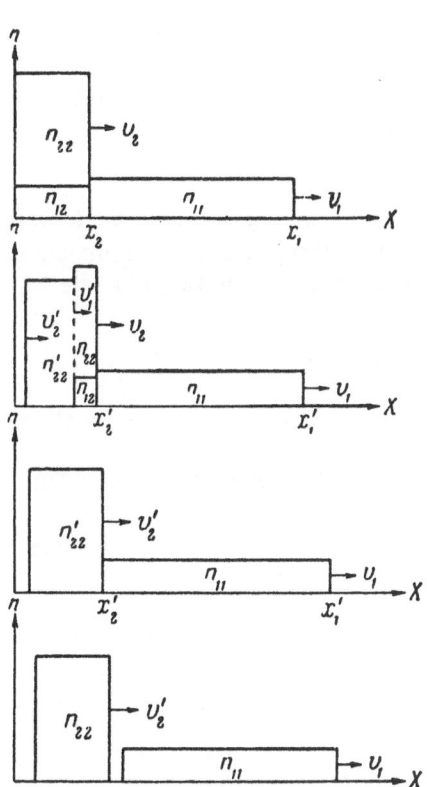

Fig. 22. Separation of two substances by elution in the case of interdependent equilibrium sorption dynamics.

Here all that was said on the question of the sorption dynamics of one substance in the case of a linear sorption isotherm is fully applicable. In the ideal case, if there were no perturbing factors and $D^* = 0$, the boundaries of the zones would possess straight breaks, and at some stage of elution, there would be an ideal complete separation of the mixture. However, as a result of the action of longitudinal and kinetic factors, in the case of linear isotherms there will be a gradual blurring of the zones. Using the formulas of sorption dynamics of one substance in the case of a linear isotherm, we can estimate the degree of mutual contamination of the zones and the efficiency of the separation.

In the case of interdependent sorption of substances, the process of elution sorption dynamics will be more complex. Hence, to elucidate the essence of the process, we assume, in the first place, that $D^* = 0$, and, in the second place, we consider the simplest case of separation of a mixture of two substances under conditions of equilibrium sorption dynamics [241].

The picture of elution sorption dynamics of two substances is shown schematically in Fig. 22.

At the initial moment we have the primary chromatogram of the two substances. From the beginning of elution, the following process takes place. The concentration of the first component $n_{1,1}$ in front of the forward front remains unchanged. Hence the front will move at a rate

$$v_1 = u \frac{n_{1,1}}{n_{1,1} + N_{1,1}}. \tag{V.16}$$

The forward front of the second component will move at a rate

$$v_2 = u \frac{n_{2,2}}{n_{2,2} + N_{2,2}}. \tag{V.17}$$

Delivery of the substances to be sorbed to the column no longer occurs at the rear front from the moment when elution begins—the solvent is delivered. The rear front of the first component in the second zone begins to move at a rate

$$v_1' = u \frac{n_{1,2}}{n_{1,2} + N_{1,2}}. \tag{V.18}$$

Since the second component is less sorbable, $v_1' > v_2$, and the zone with concentration $n_{1,2}$ gradually begins to contract. The rear front of the second component in the second zone should lag behind the rear front of the first component in the same zone during elution. But since equilibrium is established instantaneously, then a new zone of the pure second component with a new equilibrium concentration $n_{2,2}'$ is formed instantaneously behind the front of the concentration $n_{1,2}$ in the second zone. The rear front of this pure zone will already move at its own rate

$$v_2' = u \frac{n_{2,2}'}{n_{2,2}' + N_{2,2}'}. \tag{V.19}$$

At a definite moment of elution, the entire first component will pass into the first zone, and two pure, at first in contact, but at the next moment already divergent zones will arise in the column—complete separation of the mixture will occur. This moment of time can be determined from the following considerations:

Let a time t_0 be expended for the formation of the primary chromatogram. During this time, the forward front of the second component will move over a distance $x_2 = v_2 t_0$. The moment of separation of the mixture of two substances t is reached when the front of the first component with concentration $n_{1,2}$, moving at a rate v_1', reaches the forward front of the second component, moving at a rate v_2.

The coordinate corresponding to this moment t will be

$$x_2' = v_2 t_0 + v_2 t = v_1' t. \tag{V.20}$$

From the last equation, after substitution of the formulas (V.17) and (V.18), we obtain the time sought:

$$t = t_0 \left[\frac{n_{1,2}/(n_{1,2} + N_{1,2})}{n_{2,2}/(n_{2,2} + N_{2,2})} - 1 \right]^{-1}. \tag{V.21}$$

If instead of the time t_0, the width of the second zone x_2 is known, then instead of (V.20) we shall have the following equation:

$$x_2' = x_2 + v_2 t = v_1' t, \tag{V.22}$$

from which, just as with formula (V.21), we obtain:

$$t = \frac{x_2}{v_1' - v_2} = \frac{x_2}{u \left(\dfrac{n_{1,2}}{n_{1,2} + N_{1,2}} - \dfrac{n_{2,2}}{n_{2,2} + N_{2,2}} \right)}. \tag{V.23}$$

Both calculation formulas—(V.21) and (V.23)—permit the formulation of the condition of the most effective separation of the mixture of two substances (least time of separation t):

$$\frac{n_{1,2}}{n_{1,2} + N_{1,2}} \gg \frac{n_{2,2}}{n_{2,2} + N_{2,2}}, \tag{V.24}$$

from which

$$\frac{n_{1,2}}{N_{1,2}} \gg \frac{n_{2,2}}{N_{2,2}} \quad \text{or} \quad h_{1,2} > h_{2,2}. \tag{V.25}$$

If $h_{1,2} = h_{2,2}$, then $t \rightarrow \infty$, and separation is impossible.

The concentration $n'_{2,2}$ in the pure zone of the second component is determined from the equation of balance, analogously to the equations (IV.47) and (IV.49). At the rear boundary of the first component in the second zone, we have the following material balance of the second component:

$$(n'_{2,2} - n_{2,2})\, u = [(n'_{2,2} + N'_{2,2}) - (n_{2,2} + N_{2,2})]\, v'_1, \tag{V.26}$$

from which after substitution of formula (V.18) and algebraic simplification, we obtain:

$$n'_{2,2} - n_{2,2} = \frac{n_{1,2}}{N_{1,2}}(N'_{2,2} - N_{2,2}), \tag{V.27}$$

or after substitution of the sorption isotherms

$$n'_{2,2} - n_{2,2} = \frac{n_{1,2}}{N_{1,2}}[f(n'_{2,2}) - f(n_{1,2}, n_{2,2})] \tag{V.28}$$

we obtain an equation with one unknown $n'_{2,2}$.

The analysis of the case of separation of a mixture of three components or more presents difficulties only on account of the cumbersome nature of the calculations. In principle, the same method can be used here as in the case of two components of the system.

Until now we were assuming that longitudinal and kinetic effects were absent at the boundaries of the zones during elution. Under action conditions, on the other hand, blurring of the boundaries of the chromatographic zones will occur. In the case of a convex isotherm, under equilibrium and nonequilibrium conditions, the forward fronts of the zones acquire a steady-state form at the asymptotic stage of the process, while the rear fronts will be continuously blurred, forming "tails." In the case of a concave isotherm, the picture will be the opposite—the rear fronts will be steady-state, while the forward fronts will be stretched out ("tongues"). The formation of "tails" and "tongues" in elution chromatography is an extremely harmful phenomenon, which greatly reduces the effectiveness of separation of a mixture of substances.

In recent years, methods have been proposed for eliminating these harmful effects. They include the methods of the so-called gradient chromatography—gradients of the temperature field (thermal chromatography) [18, 54, 60], concentration gradients of the eluting solution [157, 203], the Spedding method [226], etc.

In spite of the substantial deviations of the real chromatographic process from the ideal process of equilibrium chromatography, the theoretical results obtained in the equilibrium theory are of great practical value. Such very important parameters of the process as the rate of motion of the fronts of the chromatographic zones, the equilibrium concentrations of the components in the zones, and the positions of the maxima of the zones can be predicted on the basis of the theory of equilibrium sorption dynamics. A knowledge of the sorption isotherms provides the possibility of predicting the system of sorption dynamics and the efficiency of the separation of mixtures.

CHAPTER VI

THEORY OF DISPLACEMENT CHROMATOGRAPHY

1. Equilibrium Displacement Sorption Dynamics in the Absence of Longitudinal

Effects ($D^* = 0$)

First of all, let us note that from the theoretical standpoint, there is essentially no difference between displacement and elution sorption dynamics. Moreover, elution sorption dynamics can be considered from the theoretical standpoint as a particular case of displacement sorption dynamics.

In elution sorption dynamics, a pure solvent enters the column after the primary frontal chromatogram has been obtained and creates conditions for the displacement and forward motion of the sorbed substances.

In displacement sorption dynamics, after the primary, frontal chromatogram has been obtained, a solution of any other substance is introduced into the column. Let us assume that the same solvent from which the primary chromatogram was obtained is used as the solvent.

The basic principles of displacement chromatography can be observed for the example of the displacement sorption dynamics of one substance. Let us assume that a primary zone with a straight break of the front (see Fig. 18) has been formed in the column under equilibrium conditions and $D^* = 0$. The process of displacement consists of introducing some component (another or even the same) into the column.

Equilibrium displacement sorption dynamics of the primary component and the displacing component will be described by the following initial and boundary conditions:

$$t = 0, \ 0 \leqslant x \leqslant x_0, \ n = n_0, \ N = N_0; \left.\begin{array}{l}\\ \\\end{array}\right\} \text{(primary zone)} \qquad \text{(VI.1)}$$
$$x > x_0, \ n = 0, \ N = 0;$$

$$t > 0, \ x = 0, \ n = n_d^0, \ N = N_d^0; \left.\begin{array}{l}\\ \\\end{array}\right\} \begin{array}{l}\text{(instantaneous sorption}\\ \text{of displacer)}\end{array} \qquad \text{(VI.2)}$$
$$x = \infty, \ n = 0, \ N = 0.$$

Hence, according to these equations, two equilibrium concentrations in the solution composition—n_0 and n_d^0—exist in the column at the initial moment. According to the Wilson law, each of these concentration points should move at its own characteristic rate, depending on the value of the distribution ratio:

$$v = u \frac{n_0}{n_0 + N_0} = u \frac{h}{1+h}; \qquad \text{(VI.3)}$$

$$v_d = u \frac{n_d^0}{n_d^0 + N_d^0} = u \frac{h_d}{1 + h_d}. \qquad \text{(VI.4)}$$

Moreover, both the forward point of the zone of the primary component and the rear point should move at the rate v. However, the behavior of the components at the boundary of their initial contact depends on the ratio of the velocities v and v_d. Three cases are possible here: $v_d < v$, $v_d = v$, and $v_d > v$. In the first case, when $v_d < v$, the rear boundary of the primary zone will move at a greater velocity, and the zone of the component introduced will immediately lag behind the zone of the primary substance, as is shown in Fig. 23.

Between the zone of the primary substance and the zone of the displacer, an interlayer of pure solvent is formed. This case cannot be called displacement, since actually we have here elution sorption dynamics.

The movement of the zone of the primary substance actually occurs under the displacing action of the solvent.

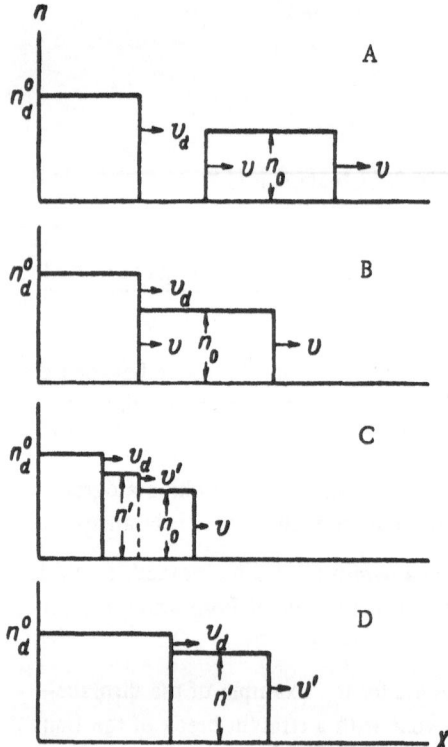

Fig. 23. Equilibrium displacement sorption dynamics in the absence of longitudinal effects. A) Condition $v_d < v$; B) condition $v_d = v$; C, D) condition $v_d > v$; establishment of a steady-state system of displacement.

The substance contained in the introduced solution forms an independent zone, the front of which moves at a rate $v_d < v$.

Hence, the presence of such a component in the solution exerts no influence on the sorption dynamics of the primary component.

An analogous picture of sorption dynamics will exist when $v_d = v$. In this case, the forward and rear boundaries of the zone of the primary component and the forward boundary of the zone of the substance introduced together with the solvent will move at the same velocities. No "gap" in the form of a zone of the pure solvent is formed between the zones. However, in this case also we shall have independent motion of the two zones. This case can also be classed as elution sorption dynamics (see Fig. 23).

The true process of displacement sorption dynamics will occur in the case when $v_d > v$. Then the rear front of the primary zone should also have moved with the velocity of the forward front of the primary zone, but this cannot occur, since the displacer component overtakes the rear front of the primary component at a greater velocity. The front of the displacing component can move at a velocity $v_d > v$ only in the case when a new zone of the primary component, the forward edge of which will move at the same velocity v_d, is formed in front of it. Figuratively speaking, the zone of the primary component should be, as it were, compressed when $v_d > v$.

In view of the equilibrium character of the sorption dynamics, a new zone of the primary component, the forward front of which will move at a velocity $v' = v_d$ (see Fig. 23), is formed instantaneously in front of the displacer, from the very beginning of the displacement process.

The intermediate zone of the primary component will gradually expand, while the primary zone will be contracted ($v' = v_d > v$), and, finally, will disappear at a definite moment.

From this moment on begins the steady-state process of motion of the primary component and forward front of the displacer at a constant velocity v_d. The time elapsed in the process of establishment of a steady-state can be calculated from the following obvious condition:

$$v't = v_d t = x_0 + vt, \tag{VI.5}$$

from which

$$t = \frac{x_0}{v_d - v} = t_0 \left(\frac{v_d}{v} - 1 \right)^{-1}, \tag{VI.6}$$

where t_0 is the time of formation of the primary zone.

The concentrations n' and N' of the primary substance in the new steady-state zone can be calculated from the following functions:

$$v_d = u \ \frac{n_d^0}{n_d^0 + N_d^0} = u \ \frac{h_d}{1 + h_d} = v' = u \ \frac{n'}{n' + N'} = u \ \frac{h'}{1 + h'}, \tag{VI.7}$$

from which

$$h' = h_d, \ n' = h_d N', \ n' = h_d f(n'). \tag{VI.8}$$

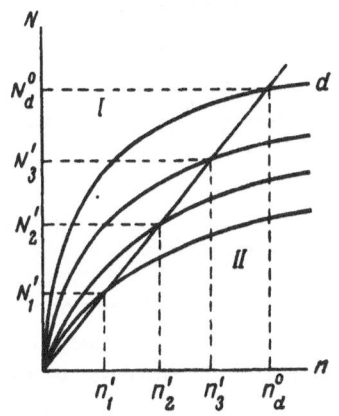

Fig. 24. For the theory of displacement sorption dynamics. Practical determination of the concentrations in the zones of the steady-state displacement chromatogram. I) Region of displacement sorption dynamics ($h_d > h$); II) region of elution sorption dynamics ($h_d < h$); d) sorption isotherm of the displacer.

The last equation will also be the equation for calculating the value of n' sought. $N' = f(n')$ can be calculated by determining it according to the sorption isotherm.

Let us consider the basic condition under which a system of displacement sorption dynamics will be set up: $v_d > v$. Let us assume that there is an infinite variety of sorption isotherms, filling the space of the coordinates (N, n). According to the boundary condition (VI.2), let us denote a point with coordinates (n_d^0, N_d^0) on the plane of the coordinates (N, n) (Fig. 24).

This should be a point lying on the sorption isotherm of the displacing component. The latter can take any form, not only for the displacer, but also for the primary component. However, the sorption isotherms should be set.

The condition of a displacement system of sorption dynamics, $v_d > v$, means that for the primary zone, the distribution ratio takes the form $h < h_d$. A straight line passing through the origin and the point (n_d^0, N_d^0) breaks up the plane of the coordinates (N, n) into two regions: above this straight line we shall have $h < h_d$, and below $h > h_d$. Thus, if for the primary zone the value of the distribution ratio $h < h_d$, and, consequently, the point (n_0, N_0) falls in the region below the straight line $N = (1/h_d)n$ and $h > h_d$, then $v_d < v$, and the system of sorption dynamics will be of an elution character.

Knowing the sorption isotherm of the primary component and its concentration in the initial solution, we can always select a displacing component with sorption isotherm known for it, and its concentration n_0 such that the condition of displacement sorption dynamics $h < h_d$ will be observed. In this case, no special conditions are set for the form of the sorption isotherms. Moreover, no conditions are set with respect to the sorbability of the displacing component in comparison with the primary component. This means that the system of displacement can be set up both in the case of greater and in the case of smaller sorbabilities of the displacer in comparison with the sorbability of the primary component.

Let us cite an example. Figure 25 presents two convex isotherms. Let the lower isotherm, reflecting lower sorbability than the upper, belong to the displacer, and the upper to the primary component. Let the concentrations in the primary zone be n_0 and N_0. Let us denote the corresponding point (n_0, N_0) on the upper

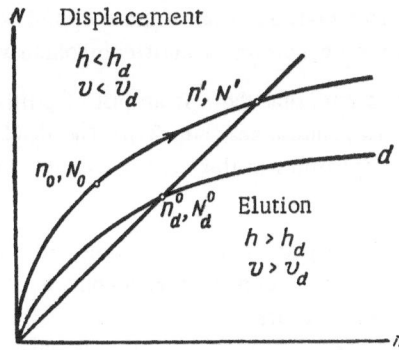

Fig. 25. Conditions of displacement of one substance by a displacer with lower sorbability for convex sorption isotherms. d) Sorption isotherms of the displacer.

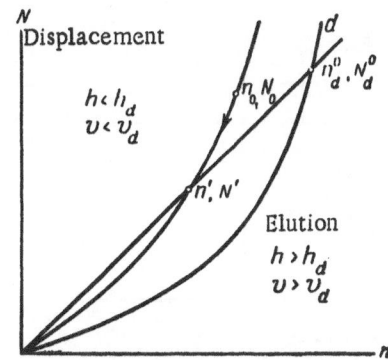

Fig. 26. Conditions of displacement of one substance by another in the case of concave sorption isotherms. d) Sorption isotherm of displacer.

isotherm. In order to set up a displacement system of sorption dynamics, a concentration of the displacer n_d^0 should be selected such that the point (n_0, N_0) will fall in the region above the straight line $N = (1/h_d)n$, passing through the point (n_d^0, N_d^0) (see Fig. 25).

It is evident from formula (VI.6) that the optimum time of establishment of a steady-state process of displacement is achieved when

$$v_d \gg v \quad \text{or} \quad \frac{n_d^0}{n_d^0 + N_d^0} \gg \frac{n_0}{n_0 + N_0} \; ; \; h \ll h_d. \qquad \text{(VI.9)}$$

As can be seen from Figs. 24 and 25, the concentration in the steady-state zone n' can be determined graphically according to the point of intersection of the straight line $N = (1/h_d)n$ with the isotherm of the primary component. Moreover, in the case of convex isotherms, the new concentration n' in the steady-state zone will be higher than the concentration of the component in the primary zone.

Thus, in this case the displacer leads to concentration, enrichment of the primary component. This result is of great practical significance. The degree of the indicated concentration of the primary component is higher, the greater the value of h_d, i.e., the higher the velocity v_d. From the law of conservation of matter for the primary component, we shall have

$$x_0 n_0 = x'n', \qquad \text{(VI.10)}$$

where x' is the width of the new steady-state zone in displacement.

The degree of concentration

$$\alpha = \frac{n'}{n_0} = \frac{x_0}{x'}. \qquad \text{(VI.11)}$$

However, the process of concentration of the substance during displacement occurs only in the case of convex sorption isotherms.

In the case of concave sorption isotherms, as is shown in Fig. 26, in the new steady-state zone the concentration n' of the primary component should be lower than in the primary zone. Thus, in this case dilution of the primary component is obtained. The cases of mixed sorption isotherms of the first component and the displacer can also be analyzed on the basis of mixed sorption isotherms of the first component and the displacer can also be analyzed on the basis of the general condition of a displacement system of sorption dynamics: $v_d > v (h_d > h)$.

Let us turn to a consideration of more complex cases of displacement sorption dynamics of multicomponent systems.

First let us assume that there is an initial primary frontal chromatogram of two substances (Fig. 27). The displacement system of sorption dynamics in this case is characterized by the condition $v_d > v_2$.

The process of displacement, beginning with the fact that a new zone of the second component with concentrations n_2' and N_2', which will move at a rate

$$v_2' = u \, \frac{n_2'}{n_2' + N_2'} = v_d = u \, \frac{n_d^0}{n_d^0 + N_d^0}. \qquad \text{(VI.12)}$$

is formed instantaneously at the rear boundary of the second primary zone as a result of the equilibrium character of the sorption dynamics.

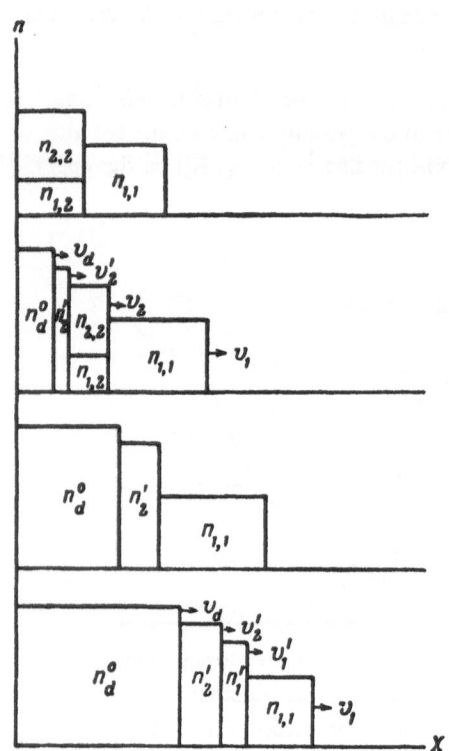

Fig. 27. Dynamics of displacement of two substances.

In this case, the second primary zone will be reduced: the first component will pass into the first zone, while the second will pass into a new zone. The forward front of the first component will move as before at a velocity $v_1 = u(n_{1,1})/(n_{1,1} + N_{1,1})$, while the forward front of the second zone will move at a velocity $v_2 = u(n_{2,2})/(n_{2,2} + N_{2,2})$.

At a definite moment of time, which can easily be calculated on the basis of a knowledge of the rates of motion v_2' and v_2, the second primary zone disappears. From this moment on, there will be a zone of the pure second component at the boundary of the first component. A new concentration of the first component n_1' is instantaneously formed, and the forward boundary of the new zone of the first component begins to move at a velocity

$$ v_1' = u \; \frac{n_1'}{n_1' + N_1'} = v_d = u \; \frac{n_d^0}{n_d^0 + N_d^0} \cdot \qquad \text{(VI.13)} $$

The new zone of the first component will expand, while the old zone will contract. At some stage of the process, this old zone of the first component disappears. From this moment on, a steady-state process of displacement of pure zones of the two components at velocities $v_1' = v_2 = v_d$ begins in the column. Thus, theoretically complete separation of the mixture of two substances should occur.

In the case of a multicomponent chromatogram of three substances or more, the process of displacement will occur analogously. The process of formation of a pure zone of the j-th component begins with the j-th, last component. Then a pure steady-state zone of the (j–1)-th component is formed, then a zone of the (j–2)-th component, and so forth—all the way to the first component.

At the steady-state stage of the process, pure zones of all the components will move in the column at velocities equal to the rate of motion of the zone of the displacer v_d (Fig. 28).

Equality of the velocities $v_1' = v_2' = \ldots = v_j' = v_d$ is equivalent to equality of the distribution ratios of the components:

$$ h_1' = h_2' = \ldots = h_j' = h_d. $$

The new established concentrations of each component can be found according to the sorption isotherm:

$$ N_i' = f(n_i'), \quad n_i' = h_d f(n_i'). \qquad \text{(VI.14)} $$

If the experimental sorption isotherms are known, then the same problem can be solved graphically, as is shown, for example, in Fig. 24. The concentrations sought are obtained by intersection of the isotherms by the straight line $N = (1/h_d)n$, which corresponds to the solution of the system of equations (VI.14).

For those components whose isotherms do not give intersection, no steady-state system of mutual displacement is established. Such substances will "run away" forward, being separated according to the principle of elution sorption dynamics.

The width of the zone at the steady-state stage of the process in displacement can be calculated after determination of the equilibrium concentrations. If an amount of the i-th component M_i is introduced into the column, then the width of the zone of this component in the displacement chromatogram will be:

$$ \delta_i = \frac{M_i}{n_i' + N_i'} \cdot \qquad \text{(VI.15)} $$

Since the quantity $n_i' + N_i'$ (height of the step) at a given h_d is a constant for each substance, then $\delta_i \sim M_i$. Thus, the width of the zone for the characteristic quantity $n_i' + N_i'$ is a measure of the amount of the substance in the initial mixture to be chromatographed.

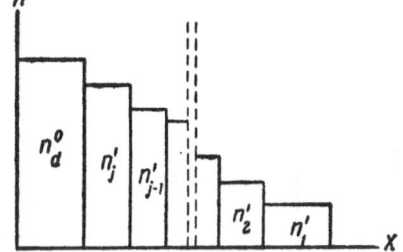

Fig. 28. Displacement chromatogram. Steady-state system of motion of the zones.

This theoretical result is utilized in the so-called Tiselius-Claesson displacement analysis [228, 77], which is used for the quantitative analysis of complex mixtures of substances.

2. Displacement Sorption Dynamics under the Action of Longitudinal and Kinetic Effects

In real chromatographic columns, as is well known, even under conditions close to equilibrium, longitudinal effects operate, blurring the boundaries ($D^* \neq 0$). Under these conditions, a vital influence on the behavior of the components at the fronts of the zones will be exerted by the type of sorption isotherm.

In the case of a displacement system of sorption dynamics, when $h_d > h$, the most favorable conditions for the process will be created in the case of a convex sorption isotherm of the displacer. In this case, independent of whether the sorption isotherm of the component to be displaced is convex or concave, a steady-state front of the displacer and the j-th component is formed at the boundary between the zone of the displacer and the j-th zone of the primary chromatogram. It can be assumed that an analogous picture will exist for other components, independent of the convexity or concavity of the sorption isotherms.

Depending on the type of isotherm, only the forward front of the first component will be stabilized or blurred. A calculation of the contour of the steady-state fronts between the zones can be performed according to the Zel'dovich-Todes method.

Since all the fronts will move at a constant rate at the steady-state stage of displacement sorption dynamics, equal to the rate of motion of the front of the displacer v_d, the system of differential equations of sorption dynamics at $D^* \neq 0$ under equilibrium conditions

$$\frac{\partial n_i}{\partial t} + u \frac{\partial n_i}{\partial x} + \frac{\partial N_i}{\partial t} = D_i^* \frac{\partial^2 n_i}{\partial x^2},$$
$$1 \leqslant i \leqslant j + 1, \quad j + 1 = d \tag{VI.16}$$

can be transformed by the substitution

$$z_d = x - v_d\, t \tag{VI.17}$$

to normal differential equations

$$-v_d \frac{dn_i}{dz_d} + u \frac{dn_i}{dz_d} - v_d \frac{dN_i}{dz_d} = D_i^* \frac{d^2 n_i}{dz_d^2}. \tag{VI.18}$$

Since each zone of components to be separated will have two fronts—forward and rear—boundary conditions must be set for both fronts:

For the forward front:

$$\left.\begin{aligned}
z_d = +\infty, \quad n_i = 0, \quad N_i = 0, \quad \frac{dn_i}{dz_d} = 0;\\
z_d = -\infty, \quad n_i = n_i', \quad N_i = N_i', \quad \frac{dn_i}{dz_d} = 0;
\end{aligned}\right\} \tag{VI.19}$$

For the rear front:

$$\left.\begin{aligned}
z_d = +\infty, \quad n_i = n_i', \quad N_i = N_i', \quad \frac{dn_i}{dz_d} = 0;\\
z_d = -\infty, \quad n_i = 0, \quad N_i = 0, \quad \frac{dn_i}{dz_d} = 0.
\end{aligned}\right\} \tag{VI.20}$$

The two boundary conditions (VI.19) and (VI.20), after substitution into equation (VI.18), give:

$$D_i^* \frac{dn_i}{dz_d} = (u - r_d)\, n_i - v_d N_i; \tag{VI.21}$$

$$v_d = u\, \frac{n_i'}{n_i' + N_i'} = u\, \frac{h_i'}{1 + h_i'} = r_i'. \tag{VI.22}$$

Analogous boundary conditions (VI.19)-(VI.20) and the equations (VI.21) and (VI.22) can be written for the $(i+1)$-th or any other component.

At the boundary between the $(i+1)$-th and i-th zones, there are two fronts: the front of the $(i+1)$-th component and the front of the i-th component. Consequently, there is a mixed two-component transition zone. The system of the following two equations should be taken for the calculation of the contours of the two fronts:

$$D_i^* \frac{dn_i}{dz_d} = (u - r_d)\, n_i - v_d N_i; \tag{VI.23}$$

$$D_{i+1}^* \frac{dn_{i+1}}{dz_d} = (u - r_d)\, n_{i+1} - r_d N_{i+1}, \tag{VI.24}$$

whereupon

$$N_i = f_i\,(n_i,\, n_{i+1}); \quad N_{i+1} = f_{i+1}\,(n_i,\, n_{i+1}).$$

The distribution of the components between the sorbent and the mobile phase is determined here by the sorption isotherm.

If sorption kinetics operate as the blurring factor of sorption dynamics, while the longitudinal effects of blurring can be neglected ($D^* = 0$), then for the transition region between the $(i+1)$-th and i-th zones, the following system of differential equations should be written:

$$\frac{\partial n_k}{\partial t} + u\, \frac{\partial n_k}{\partial x} + \frac{\partial N_k}{\partial t} = 0; \tag{VI.25}$$

$$\frac{\partial N_k}{\partial t} = \psi_k\,(n_i,\, n_{i+1};\, N_i,\, N_{i+1}); \tag{VI.26}$$

$$k = i \quad \text{or} \quad k = i+1,\ 1 \leqslant i \leqslant j+1.$$

The contours of the steady-state fronts can be calculated according to the Zel'dovich-Todes method. For this let us introduce the already known substitution (VI.17) and transform the system (VI.25), (VI.26):

$$-v_d \frac{dn_k}{dz_d} + u\, \frac{dn_k}{dz_d} - v_d \frac{dN_k}{dz_d} = 0; \tag{VI.27}$$

$$-v_d \frac{dN_k}{dz_d} = \psi_k\,(n_i,\, n_{i+1};\, N_i,\, N_{i+1}). \tag{VI.28}$$

Applying the boundary conditions (VI.19)-(VI.20) to the $(i+1)$-th and i-th components, and substituting them into the equation obtained after integration of equation (VI.27), we shall have:

$$\left.\begin{aligned}
(u - v_d)\, n_i - v_d N_i &= 0, \\
(u - v_d)\, n_i' - v_d N_i' &= 0, \\
(u - v_d)\, n_{i+1} - v_d N_{i+1} &= 0, \\
(u - v_d)\, n_{i+1}' - v_d N_{i+1}' &= 0.
\end{aligned}\right\} \tag{VI.29}$$

Using analogous conditions for the displacer, we also obtain

$$\left.\begin{aligned} (u - v_d)\, n_d - v_d N_d = 0, \\ (u - v_d)\, n_d^0 - v_d N_d^0 = 0. \end{aligned}\right\} \tag{VI.30}$$

From all these equations, it follows that

$$v_d = u\,\frac{n_i}{n_i + N_i} = u\,\frac{n_i'}{n_i' + N_i'} = u\,\frac{n_{i+1}}{n_{i+1} + N_{i+1}} = u\,\frac{n_{i+1}'}{n_{i+1}' + N_{i+1}'} = u\,\frac{n_d}{n_d + N_d} = u\,\frac{n_d^0}{n_d^0 + N_d^0}, \tag{VI.31}$$

from which the following ratios are established among the nonequilibrium and equilibrium concentrations of the components in the fronts:

$$\frac{n_i}{N_i} = \frac{n_i'}{N_i'} = \frac{n_{i+1}}{N_{i+1}} = \frac{n_{i+1}'}{N_{i+1}'} = \frac{n_d}{N_d} = \frac{n_d^0}{N_d^0} = h_d = \text{const.} \tag{VI.32}$$

This means that for displacement sorption dynamics under nonequilibrium conditions, when $D^* = 0$, a linear relationship is observed in the steady-state fronts between the nonequilibrium concentrations contained in the mobile phase and in the sorbent, i.e., a linear Zel'dovich function is observed.

The contours of the fronts can be calculated from the equations of sorption kinetics (VI.28). For example, for the fronts of the transition region between the $(i + 1)$-th and i-th zones, we shall have the following system of equations:

$$\left.\begin{aligned} - v_d\,\frac{dn_i}{dz_d} = \psi_i\,(n_i,\, n_{i+1};\; N_i,\, N_{i+1}), \\ - v_d\,\frac{dn_{i+1}}{dz_d} = \psi_{i+1}\,(n_i,\, n_{i+1};\; N_i,\, N_{i+1}). \end{aligned}\right\} \tag{VI.33}$$

This system, together with formulas (VI.31) and (VI.32) should give the distribution of substances in the steady-state fronts sought.

LITERATURE CITED

1. Adam, N. K. Surface Physics and Chemistry (Moscow–Leningrad, Gostekhizdat, 1947).
2. Akopyan, A. A. Chemical Thermodynamics (Moscow, Vysshaya Shkola, 1963).
3. Anokhin, V. L. Zhurnal Fizicheskoi Khimii, 31:976 (1957).
4. Apel'tsin, I. É., Klyachko, V. A., Lur'e, Yu. Yu., and Smirnov, A. S. Ion Exchangers and Their Application (Moscow, Standartgiz, 1949).
5. Belen'kaya, I. M. Dissertation (Moscow, 1949).
6. Bikson, Ya. M. Dissertation (Moscow, 1950).
7. Bikson, Ya. M. Zhurnal Fizicheskoi Khimii, 27:1530 (1953).
8. Bikson, Ya. M. Zhurnal Fizicheskoi Khimii, 28:1017 (1954).
9. Block, R., LeStrange, R., and Zweig, G. Paper Chromatography [Russian translation] (Moscow, IL, 1954).
10. Bonch-Bruevich, V. A. Uspekhi Fizicheskikh Nauk, 40:369 (1950).
11. Bresler, S. E. Doklady Akademii Nauk SSSR, 90:205 (1953).
12. Bresler, S. E., and Uflyand, Ya. S. Zhurnal Tekhnicheskoi Fiziki, 28:1443 (1953).
13. Bresler, S. E. Doklady Akademii Nauk SSSR, 97:699 (1954).
14. Brunauer, S. Adsorption of Gases and Vapors [Russian translation] (Moscow, IL, 1948).
15. Budak, B. M., Samarskii, A. S., and Tikhonov, A. N. Collection of Problems on Mathematical Physics (Moscow, Gos. Izd. Tekhn.-Teor. Lit., 1956).
16. Bunge, N. A. Chemical Engineering of Water (Kiev, 1878).
17. Barrer, R. Diffusion in Solids [Russian translation] (Moscow, IL, 1948).
18. Vagin, E. V., and Zhukhovitskii, A. A. Doklady Akademii Nauk SSSR, 94:273 (1954).
19. Vetrov, B. N. Dissertation (Leningrad, 1952).
20. Vetrov, B. N., and Todes, O. M. Zhurnal Technicheskoi Fiziki, 25:1217 (1955).
21. Vetrov, B. N., and Todes, O. M. Zhurnal Tekhnicheskoi Fiziki, 25:1232 (1955).
22. Vetrov, B. N., and Todes, O. M. Zhurnal Tekhnicheskoi Fiziki, 25:1242 (1955).
23. Wigner, G. Selected Works. Physicochemical Investigations of Soils [Russian translation] (Moscow, Sel'khozgiz, 1941).
24. Voznesenskii, S. A. Physicochemical Processes of Water Pruification (Moscow–Leningrad, Gosstroiizdat, 1934).
25. Vol'kenshtein, F. F. Zhurnal Fizicheskoi Khimii, 26:1462 (1952).
26. Gapon, E. N., Gapon, T. B., and Shemyakin, F. M. Doklady Akademii Nauk SSSR, 58:597 (1947).
27. Gapon, E. N., and Gapon, T. B. Doklady Akademii Nauk SSSR, 59:921 (1948).
28. Gapon, E. N., and Gapon, T. B. Doklady Akademii Nauk SSSR, 60:401 (1948).
29. Gapon, E. N., and Gapon, T. B. Doklady Akademii Nauk SSSR, 60:817 (1948).
30. Gapon, E. N., and Gapon, T. B. Uspekhi Khimii, 17:452 (1948).
31. Gapon, E. N., and Gapon, T. B. Zhurnal Prikladnoi Fiziki, 21:937 (1948).
32. Gapon, E. N., and Gapon, T. B. Zhurnal Fizicheskoi Khimii, 22:859 (1948).
33. Gapon, E. N., and Gapon, T. B. Zhurnal Fizicheskoi Khimii, 22:979 (1948).
34. Gapon, E. N., and Gapon, T. B. M. S. Tsvet's Chromatographic Analysis and Ion Exchange; see [155, p. 9].
35. Gapon, E. N., and Gapon, T. B. Zhurnal Obshchei Khimii, 19:1627 (1949).
36. Gapon, E. N., Gapon, T. B., and Zhupakhina, E. S. Theory of Ion Exchange Chromatography; see [70, p. 5].
37. Gapon, E. N., and Belen'kaya, I. M. Kolloidnyi Zhurnal, 14:323 (1952).
38. Gedroits, K. K. Theory of the Absorption Capacity of Soils (Moscow, Sel'khozgiz, 1932).
39. Gurvich, A. M., and Gapon, T. B. A Method of Adsorption-Complex-Forming Chromatography; see [156, p. 355].

40. Danilevskii, A. I. Arch. Path. Anat. u. Physiol. u. Klin. Med., 25:279 (1869).

41. Dubinin, M. M. Zhurnal Russkogo Fiziko-khimicheskogo Obshchestva, 62:683 (1930).

42. Dubinin, M. M., Parshin, S. L, and Pupyrev, A. A. Zhurnal Russkogo Fiziko-khimicheskogo Obsh-chestva, 62:1947 (1930).

43. Dubinin, M. M. Zhurnal Prikladnoi Khimii, 4:283 (1931).

44. Dubinin, M. M. Physicochemical Bases of the Sorption Technique (Moscow—Leningrad, ONTI, 1935).

45. Dubinin, M. M., and Khrenova, M. V. Zhurnal Prikladnoi Khimii, 9:1204 (1936).

46. Dubinin, M. M., and Yavich, S. Zhurnal Prikladnoi Khimii, 9:1191 (1936).

47. Dubinin, M. M., and Chmutov, K. V. Physicochemical Bases of the Gas Mask (Moscow, Izd. Voenn. Akad. Khim. Zashch., 1939).

48. D'yachkovskii, S. L Zhurnal Obshchei Khimii, 1:81 (1931).

49. D'yachkovskii, S. L Zhurnal Obshchei Khimii, 3:478 (1933).

50. D'yachkovskii, S. L Kolloidnyi Zhurnal, 12:112 (1950).

51. Zhukhovitskii, A. A. Adsorption of Gases and Vapors (Moscow, GONTI, 1938).

52. Zhukhovitskii, A. A., Zabezhinskii, Ya. L., and Saminskii, D. S. Zhurnal Fizicheskoi Khimii, 13:303 (1939).

53. Zhukhovitskii, A. A., Zabezhinskii, Ya. L., and Tikhonov, A. N. Zhurnal Fizicheskoi Khimii, 19:253 (1945).

54. Zhukhovitskii, A. A., Zolotareva, O. V., Sokolov, V. A., and Turkel'taub, N. M. Doklady Akademii Nauk SSSR, 77:435 (1951).

55. Zhukhovitskii, A. A., Turkel'taub, N. M., and Sokolov, V. A. Doklady Akademii Nauk SSSR, 88:859 (1953).

56. Zhukhovitskii, A. A., Turkel'taub, N. M., and Georgievskaya, T. V. Doklady Akademii Nauk SSSR, 92:987 (1953).

57. Zhukhovitskii, A. A., and Turkel'taub, N. M. Doklady Akademii Nauk SSSR, 94:77 (1954).

58. Zhukhovitskii, A. A., Turkel'taub, N. M., Vagin, E. V., and Shvartsman, V. P. Doklady Akademii Nauk SSSR, 96:303 (1954).

59. Zhukhovitskii, A. A. Development and Theory of the Adsorption Method of M. S. Tsvet; see [145, p. 33].

60. Zhukhovitskii, A. A., and Turkel'taub, N. M. Gas Chromatography (Moscow, Gostoptekhizdat, 1962).

61. Zabezhinskii, Ya. L., Zhukhovitskii, A. A., and Tikhonov, A. N. Zhurnal Fizicheskoi Khimii, 23:192 (1949).

62. Zelinskii, N. D., and Sadikov, V. S. Coal as a Means of Combatting Suffocating and Poisonous Gases. Experimental Investigations of 1915-1916 (Moscow, Izd. AN SSSR, 1941).

63. Zel'dovich, Ya. B. Cited in [115, 139].

64. Zimin, N. I. Transactions of the Ninth Water Supply Congress (Tiflis, 1909).

65. Ivanenko, D. D., Rachinskii, V. V., Gapon, T. B., and Gapon, E. N. Doklady Akademii Nauk SSSR, 60:1189 (1948).

66. Il'in, B. V. Nature of Adsorption Forces (Moscow, Gosteortekhizdat, 1952).

67. Collection: Ion Exchange [Russian translation] (Moscow, IL, 1951).

68. Collection: Ion Exchange and Its Application (Moscow, Izd. AN SSSR, 1959).

69. Collection: Ion Exchange Sorbents in Industry (Moscow, Izd. AN SSSR, 1963).

70. Collection: Research in the Field of Chromatography (Moscow, Izd. AN SSSR, 1952).

71. Collection: Research in the Field of Ion Exchange Chromatography (Moscow, Izd. AN SSSR, 1957).

72. Collection: Research in the Field of Ion Exchange, Partition, and Precipitation Chromatography (Moscow, Izd. AN SSSR, 1959).

73. Collection: Research in the Field of the Industrial Application of Sorbents (Moscow, Izd. AN SSSR, (1961).

74. Kafarov, V. V. Mass Transfer (Moscow, Vysshaya Shkola, 1962).

75. Kvitka, S. K. Testimony of the Baku Technical Committee, June 17, 1960; see [78].

76. Keilemans, A. Gas Chromatography [Russian translation] (Moscow, IL, 1959).

77. Claesson, S. Adsorption Analysis of Mixtures [Russian translation] (Moscow—Leningrad, Goskhimizdat, 1950).

78. Koshtoyants, Kh. S., and Kalymkov, K. F. Biokhimiya, 16:479 (1951).

79. Kunin, R., and Myers, R. Ion Exchange Resins [Russian translation] (Moscow, IL, 1952).

80. Courant, R., and Hilbert, D. Methods of Mathematical Analysis [Russian translation] Vols. I and II (Moscow, Gostekhizdat, 1951).

81. Kutateladze, S. S., and Borishanskii, V. M. Handbook on Heat Transfer (Moscow, Gosénergoizdat, 1959).

82. Landau, L. D., and Lifshits, E. M. Solid-State Mechanics (Moscow, Gostekhizdat, 1944).

83. Levich, V. G. Physicochemical Hydrodynamics (Moscow, Fizmatgiz, 1959).

84. Leibenzon, L. S. Motion of Natural Liquids and Gases in a Porous Medium (Moscow, Gostekhizdat, 1947).

85. Lepin', L. K., and Voznesenskii, S. A. Zhurnal Obshchei Khimii, 1:233 (1931).

86. Lipatov, S. M. Zhurnal Russkogo Fiziko-khimicheskogo Obshchestva, Chast' Khimicheskaya, 58:983 (1926).

87. Lipatov, S. M. Zhurnal Russkogo Fiziko-khimicheskogo Obshchestva, 62:1985 (1930).

88. Lovits, T. E. Selected Works on Chemistry and Chemical Engineering (Moscow, Izd. AN SSSR, 1955).

89. Lykov, A. V. Transport Phenomena in Capillary-Porous Solids (Moscow, GITTL, 1954).

90. McBain, J. Sorption of Gases and Vapors by Solids [Russian translation] (Leningrad, ONTI, Khimteoret, 1934).

91. Matveev, N. M. Methods of Integration of Normal Differential Equations (Moscow, Vysshaya Shkola, 1963).

92. Mayer, S. W., and Tompkins, E. Theoretical Analysis of the Process of Decomposition in a Column [Russian translation]; see [155, p. 212].

93. Mecklenburg, V. B. Zhurnal Russkogo Fiziko-khimicheskogo Obshchestva, 62:1723 (1930).

94. Meleshko, V. P., and Voitovich, V. B. Doklady Akademii Nauk SSSR, 102:965 (1955).

95. Meleshko, V. P. Trudy Voronezhskoi Gosudarstvennoi Universiteta, 49:55 (1958).

96. Meleshko, V. P. Dissertation (Moscow, 1962).

97. Myasnikov, L. A., and Gol'bert, K. A. Zhurnal Fizicheskoi Khimii, 27:1311 (1953).

98. Nemytskii, V. V., and Stepanov, V. V. Qualitative Theory of Differential Equations (Moscow, GITTL, 1949).

99. Ol'shanova, K. M., Kopylova, V. D., and Morozova, N. M. Precipitation Chromatography (Izd. AN SSSR, 1963).

100. Oppokov, G. V. Numerical Integration of Differential Equations in Partial Derivatives (Moscow, GTTI, 1932).

101. Panov, D. Yu. Handbook on the Numerical Solution of Differential Equations in Partial Derivatives (Moscow, GITTL, 1951).

102. Petrovskii, I. G. Readings on Equations with Partial Derivatives (Moscow—Leningrad, Gostekhizdat, 1953).

103. Collection: Surface Chemical Compounds and Their Role in Adsorption Phenomena (Izd. MGU, 1957).

104. Collection: Production, Structure, and Properties of Sorbents (Moscow, Goskhimizdat, 1959).

105. Radushkevich, L. V. Doklady Akademii Nauk SSSR, 57:471 (1947).

106. Raidil, É. K. Chemistry of Surface Phenomena (Leningrad, ONTI, 1936).

107. Rachinskii, V. V., and Gapon, T. B. Chromatography in Biology (Moscow, Izd. AN SSSR, 1953).

108. Rachinskii, V. V. Doklady Akademii Nauk SSSR, 88:701 (1953).

109. Rachinskii, V. V. Doklady Akademii Nauk SSSR, 88:883 (1953).

110. Rachinskii, V. V. Izvestiya Timiryazevskoi Sel'sko-Khozyaistvennoi Akademii, 2(3):193 (1953).

111. Rachinskii, V. V. Trudy Komis. po Analit. Khimii AN SSSR, USSR, 6(9):21 (1955).

112. Rachinskii, V. V. Zhurnal Fizicheskoi Khimii, 30:407 (1956).

113. Rachinskii, V. V. Zhurnal Fizicheskoi Khimii, 31:444 (1957).

114. Rachinskii, V. V. Zhurnal Fizicheskoi Khimii, 36:2018 (1962).

115. Rachinskii, V. V. Layer-by-Layer Method of Approximate Calculation of Chromatograms; see [71, p. 5].

116. Rachinskii, V. V. On the Theory of the Steady-State Front of Dynamic Sorption; see [72, p. 24].

117. Rachinskii, V. V. Izvestiya Timiryazevskoi Sel'sko-Khozyaistvennoi Akademii, 4(29):187 (1959).

118. Rachinskii, V. V. Izvestiya Timiryazevskoi Sel'sko-Khozyaistvennoi Akademii, 6(31):201 (1959).

119. Rachinskii, V. V. Izvestiya Timiryazevskoi Sel'sko-Khozyaistvennoi Akademii, 2(33):157 (1960).

120. Rachinskii, V. V. Izvestiya Timiryazevskoi Sel'sko-Khozyaistvennoi Akademii, 5(36):184 (1960).

121. Rachinskii, V. V. Izvestiya Timiryazevskoi Sel'sko-Khozyaistvennoi Akademii, 3(40):177 (1961).

122. Rachinskii, V. V. Izvestiya Timiryazevskoi Sel'sko-Khozyaistvennoi Akademii, 6(43):214 (1961).

123. Roginskii, S. Z. Adsorption and Catalysis on Inhomogeneous Surfaces (Moscow, Izd. AN SSSR, 1948).

124. Rozen, B. Ya. Uspekhi Khimii, 21:508 (1952).

125. Ryumpler, A. Sugar Industry (Petrograd, 1923).

126. Saldadze, K. M. Ion Exchange Resins (Moscow, Izd. AN SSSR, 1959).

127. Saldadze, K. M., Pashkov, A. B., and Titov, V. S. High-Molecular Ion Exchange Compounds (Moscow, Goskhimizdat, 1960).

128. Samsonov, G. V. Chromatography. Applications in Biochemistry (Moscow, Medgiz, 1955).

129. Samuelson, O. Applications of Ion Exchange in Analytical Chemistry [Russian translation] (Moscow, IL, 1955).

130. Sveshnikov, B. Ya. Priroda, 9:65 (1951).

131. Senyavin, M. M. Elements of the Theory of Ion Exchange and Ion Exchange Chromatography; see [68, p. 84].

132. Serpionova, E. N. Industrial Adsorption of Gases and Vapors (Moscow, Goskhimizdat, 1956).

133. Smirnov, V. I. Course in Higher Mathematics, Vols. II, III, IV (Moscow—Leningrad, Gostekhizdat, 1949-1951).

134. Tananaev, N. A. The Drop Method (Sverdlovsk-Moscow, GONTI, 1939).

135. Collection: Theory and Practice of the Use of Ion Exchange Materials (Moscow, Izd. AN SSSR, 1955).

136. Timofeev, D. P. Kinetics of Adsorption (Moscow, Izd. AN SSSR, 1962).

137. Tikhonov, A. N., Zhukhovitskii, A. A., and Zabezhinskii, Ya. L. Zhurnal Fizicheskoi Khimii, 20:113 (1946).

138. Tikhonov, A. N., and Samarskii, A. A. Equations of Mathematical Physics (Moscow, Gostekhizdat, 1953).

139. Todes, O. M. Zhurnal Prikladnoi Khimii, 18:591 (1945).

140. Todes, O. M., and Bikson, Ya. M. Doklady Akademii Nauk SSSR, 75:727 (1950).

141. Todes. O. M. Basic Problems of Adsorption and Catalysis on Moving and Suspended Layers; in Collection: Methods and Processes of Chemical Engineering (Moscow—Leningrad, Izd. AN SSSR, 1955), p. 65.

142. Todes, O. M., and Rachinskii, V. V. Zhurnal Fizicheskoi Khimii, 29:1591 (1955).

143. Todes, O. M., and Rachinskii, V. V. Zhurnal Fizicheskoi Khimii, 29:1909 (1955).

144. Todes, O. M., and Lezin, Yu. S. Doklady Akademii Nauk SSSR, 106:307 (1956).

145. Trudy Komis. po Analit. Khimii AN SSSR, 6(9) (1955).

146. Tunitskii, N. N., and Cherneva, E. P. Zhurnal Fizicheskoi Khimii, 24:1350 (1950).

147. Tunitskii, N. N., and Shenderovich, L. M. Doklady Akademii Nauk SSSR, 81:649 (1951).

148. Tunitskii, N. N., and Shenderovich, L. M. Zhurnal Fizicheskoi Khimii, 26:1423 (1952).

149. Tunitskii, N. N., Cherneva, E. P., and Andreev, V. I. Zhurnal Fizicheskoi Khimii, 28:2006 (1954).

150. Tunitskii, N. N. Doklady Akademii Nauk SSSR, 99:577 (1954).

151. Feigl, F. Spot Tests [Russian translation] (Moscow, ONTI, 1937).

152. Phillips, K. Gas Chromatography [Russian translation] (Moscow, IL, 1958).

153. Collection: Chromatography [Russian translation] (Moscow, IL, 1949).

154. Collection: Chromatography (Izd. LGU, 1956).

155. Collection: Chromatographic Method of Separating Ions [Russian translation] (Moscow, IL, 1949).

156. Collection: Chromatography, Its Theory and Applications (Moscow, Izd. AN SSSR, 1960).

157. Collection: Paper Chromatography [Russian translation] Editors L. M. Hais and K. Macek (Moscow, IL, 1963).

158. Tsvet, M. S. Trudy Varshavskogo Obshchestva Estestvoispytatelei. Otdel Biologii, 14:20 (1903).

159. Tsvet, M. S. Chromophylls in the Plant and Animal World (Warsaw, 1910).

160. Tsvet, M. S. Chromatographic Adsorption Analysis, Selected Works (Moscow, Izd. AN SSSR, 1946).

161. Chmutov, K. V. Chromatography (Moscow, Izd. AN SSSR, 1962).

162. Chudnovskii, A. F. Heat Exchange in Dispersed Media (Moscow, Gostekhizdat, 1954).

163. Shevelev, Ya. V. Dissertation (Moscow, 1956).

164. Shevelev, Ya. V. Zhurnal Fizicheskoi Khimii, 31:960 (1957).

165. Shevelev, Ya. V. Zhurnal Fizicheskoi Khimii, 31:1210 (1957).

166. Shemyakin, F. M. Physicochemical Periodic Processes (Moscow, Izd. AN SSSR, 1938).

167. Shemyakin, F. M., Mitselovskii, É. S., and Romanov, D. V. Chromatographic Analysis (Moscow, Goskhimizdat, 1955).

168. Shilov, N. A., Lepin', L. K., and Voznesenskii, S. A. Zhurnal Russkogo Fiziko-khimicheskogo Obshchestva, 61:1107 (1929).

169. Yanovskii, M. I. Doklady Akademii Nauk SSSR, 69:655 (1949).

170. Adams, V. A., and Holmes, R. L. J. Soc. Chem. Ind., 54:1 (1935).

171. Allander, C. G. Acta Polytechnica (Sweden), Chemistry Series (Including Metallurgy), 3:130 (1953).

172. Anzelius, A. Z. Angew. Math. Mech., 6:291 (1926).

173. Brimley, R. C., and Barrett, F. C. Practical Chromatography (London, 1953).

174. Cassidy, H. G. Fundamentals of Chromatography (New York, Interscience, 1957).

175. Chromatographic Analysis, Discussions of the Faraday Society, No. 7 (1949).

176. Cohn, A. Z. Wiss. Zool., 3:264 (1851); Arch. fur Mikr. Anatomie, 3:1 (1867).

177. Consden, R., Gordon, A., and Martin, A. Biochem. J., 38:224 (1944).

178. Day, D. Proc. Am. Phil. Soc., 36:112 (1897).

179. Eichgorn, E. Pogg. Ann., 105:126 (1958); Landw. Jahrb., 4:1 (1875).

180. Ekedahl, E., and Sillen, L. G. Arkiv for Kemi, Min. o. Geol., 25A:1 (1947).

181. Engler, C., and Albrecht, E. Z. Angew. Chem., 1:1889 (1901).

182. Furnas, C. C. Ind. Eng. Chem., 22:72 (1930); Bureau of Mines Bulletin No. 362 (1932).

183. Gans, R. German Patent No. 197111 (1906); U. S. Patent No. 914405 (1909).

184. Gilpin, J., and Gram, M. Am. Chem. J., 40:495 (1908).

185. Gilpin, J., and Scheeberger, P. Am. Chem. J., 50:59 (1913).

186. Glueckauf, E. Nature, 156:748 (1945).

187. Glueckauf, E. Nature, 160:301 (1947).

188. Glueckauf, E. Proc. Roy. Soc., A186:35 (1946).

189. Glueckauf, E. J. Chem. Soc., p. 1302 (1947).

190. Glueckauf, E. J. Chem. Soc., p. 1315 (1947).

191. Glueckauf, E., and Coates, J. J. J. Chem. Soc., p. 1315 (1947).

192. Glueckauf, E. J. Chem. Soc., p. 1321 (1947).

193. Glueckauf, E. J. Chem. Soc., p. 3260 (1949).

194. Glueckauf, E. Disc. Farad. Soc., No. 7, p. 11 (1949).

195. Glueckauf, E., Borker, K. H., and Kitt, G. P. Disc. Farad. Soc., No. 7, 199 (1949).

196. Glueckauf, E. Trans. Farad. Soc., 51:34 (1955).

197. Glueckauf, E. Trans. Farad. Soc., 51:1540 (1955).

198. Göppelsröder, F. Uber Kapilar-Analyse und ihre verschiedenen Anwendungen, sowie uber das Emporsteigen der Farbstoffe in den Pflanzen (Vienna, 1889).

199. Griessbach, R. Austauschadsorption in Theorie und Praxis (Berlin, 1957).

200. Harm, I. German Patent No. 95447 (1896).

201. Holmes, E. L. British Patent No. 472404 (1937).

202. Johansson, C. H., Persson, G., and Svensson, B. Sorption in Flow Through a Granular Layer (Stockholm, 1948).

203. Lederer, E., and Lederer, M. Chromatography. A Review of Principles and Applications (Amsterdam, Elsevier, 1953).

204. Liebig, J. Die Chemie in ihrer Anwendung auf Agrikultur und Physiologie (Braunschweig, 1840).

205. Liebig, J. Am. Chem. Pharm., 94:373 (1855).

206. Mantell, C. L. Adsorption (New York, London, 1945).

207. Martin, A. I. P., and Synge, R. L. M. Biochem. J., 35(12):1358 (1941).

208. Mecklenburg, W., and Kubelka, P. Z. Electrochem., 31:488 (1925).

209. Muhlfart. Ann. Phys., 3:328 (1900).

210. Nusselt, W. Zeitsch, V. D. I., 71(3):85 (1927).

211. Offord, A. C., and Weiss, J. Nature, 155:725 (1946).

212. Offord, A. C., and Weiss, J. Disc. Farad. Soc., No. 7, p. 26 (1949).

213. Opienska-Blauth, J., Waksmundski, A., and Kanski, M. Chromatografija (Warsaw, PWN, 1958).

214. Patten, H. E. U.S. Dept. Agr. Bureau of Soils Bulletin No. 52 (1958).

215. Poggendorf, S. Am. Phys. Chem., 15:126 (1858).

216. Runge, F. F. Der Bildungstrieb der Stoffe (Oranienburg, 1855).

217. Schönbein, F. Verhandlungen der Naturforsch. Ges. zu Basel., Part II, 111(249) (1861).

218. Schumann, T. E. W. J. Franklin Inst., 208:405 (1928).

219. Schwab, G. Ang. Chem., 50:546 (1937); 50:691 (1937); 51:709 (1938); 52:666 (1939); 53:39 (1940).

220. Sillen, L. G. Arkov for Kemi, Min. o. Geol., 22A(15):1 (1945).

221. Sillen, L. G., and Ekedahl, E. Arkiv for Kemi, Min. o. Geol., 22A(16):1 (1946).

222. Sillen, L. G. Arkiv for Kemi, 2(34):477 (1950).

223. Sillen, L. G. Arkiv for Kemi, 2(35):499 (1950).

224. Sillen, L. G. Nature, 166:722 (1950).

225. Smit, W. M. Disc. Farad. Soc., No. 7, p. 38 (1949).

226. Spedding, F. H., Voigt, A. F., Gladrow, E. M., and Steight, N. R. J.Am.Chem.Soc. 69:2777 (1947).

227. Thomas, H. C. J. Chem. Soc., p. 1663 (1944).

228. Tiselius, A. Arkiv for Kemi, Min. o. Geol., 14B(22):1 (1940).

229. De Vault, D. J. Am. Chem. Soc., 65:532 (1943).

230. Walter, J. E. J. Chem. Phys., 13(6):229 (1945).

231. Way, L T. J. Roy. Agr. Soc., 11:313 (1850); 15:135 (1854).

232. Weiss, J. J. Chem. Soc., 145:297 (1943).

233. Weyde, E., and Wicke, E. Koll. Z., 90(2):156 (1940).

234. Wicke, E. Koll. Z., 86(2):167 (1939).

235. Wicke, E. Koll. Z., 86(3):295 (1939).

236. Wicke, E. Koll. Z., 93(2):129 (1940).

237. Wicke, E., and Kallenbach, R. Koll. Z., 97(2):135 (1941).

238. Williams, T. I. An Introduction to Chromatography (London, 1947).

239. Williams, T. L, and Weil, H. Nature, 170:503 (1952).

240. Williams, T. I. The Elements of Chromatography (Glasgow, 1954).

241. Wilson, J. N. J. Am. Chem. Soc., 62:1583 (1940).